Neltje Blanchan

Bird Neighbors

An Introductory Acquaintance with one Hundred and Fifty Birds....

Neltje Blanchan

Bird Neighbors

An Introductory Acquaintance with one Hundred and Fifty Birds....

ISBN/EAN: 9783337059804

Printed in Europe, USA, Canada, Australia, Japan

Cover: Foto ©berggeist007 / pixelio.de

More available books at **www.hansebooks.com**

GOLDFINCH.
¾ Life-size.

BIRD NEIGHBORS. AN INTRODUCTORY ACQUAINTANCE WITH ONE HUNDRED AND FIFTY BIRDS COMMONLY FOUND IN THE GARDENS, MEADOWS, AND WOODS ABOUT OUR HOMES

BY

NELTJE BLANCHAN

WITH INTRODUCTION BY

JOHN BURROUGHS

AND FIFTY-TWO COLORED PLATES

[TWENTY-THIRD THOUSAND]

NEW YORK

DOUBLEDAY & McCLURE CO.

1899

נדפס בדפוס חורב

TABLE OF CONTENTS

INTRODUCTION

I WRITE these few introductory sentences to this volume only to second so worthy an attempt to quicken and enlarge the general interest in our birds. The book itself is merely an introduction, and is only designed to place a few clews in the reader's hands which he himself or herself is to follow up. I can say that it is reliable and is written in a vivacious strain and by a real bird lover, and should prove a help and a stimulus to any one who seeks by the aid of its pages to become better acquainted with our songsters. The pictures, with a few exceptions, are remarkably good and accurate, and these, with the various grouping of the birds according to color, season, habitat, etc., ought to render the identification of the birds, with no other weapon than an opera glass, an easy matter.

When I began the study of the birds I had access to a copy of Audubon, which greatly stimulated my interest in the pursuit, but I did not have the opera glass, and I could not take Audubon with me on my walks, as the reader may this volume, and he will find these colored plates quite as helpful as those of Audubon or Wilson.

But you do not want to make out your bird the first time; the book or your friend must not make the problem too easy for you. You must go again and again, and see and hear your bird under varying conditions and get a good hold of several of its characteristic traits. Things easily learned are apt to be easily forgotten. Some ladies, beginning the study of birds, once wrote to me, asking if I would not please come and help them, and set them right about certain birds in dispute. I replied that that would be getting their knowledge too easily; that what I and any one else told them they would be very apt to forget, but that the things they found out themselves they would always remember. We must in a way earn what we have or keep. Only thus does it become *ours*, a real part of us.

Not very long afterward I had the pleasure of walking with one of the ladies, and I found her eye and ear quite as sharp as my own, and that she was in a fair way to conquer the bird kingdom without any outside help. She said that the groves and fields, through which she used to walk with only a languid interest, were now completely transformed to her and afforded her the keenest pleasure; a whole new world of interest had been disclosed to her; she felt as if she was constantly on the eve of some new discovery; the next turn in the path might reveal to her a new warbler or a new vireo. I remember the thrill she seemed to experience when I called her attention to a purple finch singing in the tree-tops in front of her house, a rare visitant she had not before heard. The thrill would of course have been greater had she identified the bird without my aid. One would rather bag one's own game, whether it be with a bullet or an eyebeam.

The experience of this lady is the experience of all in whom is kindled this bird enthusiasm. A new interest is added to life; one more resource against ennui and stagnation. If you have only a city yard with a few sickly trees in it, you will find great delight in noting the numerous stragglers from the great army of spring and autumn migrants that find their way there. If you live in the country, it is as if new eyes and new ears were given you, with a correspondingly increased capacity for rural enjoyment.

The birds link themselves to your memory of seasons and places, so that a song, a call, a gleam of color, set going a sequence of delightful reminiscences in your mind. When a solitary great Carolina wren came one August day and took up its abode near me and sang and called and warbled as I had heard it long before on the Potomac, how it brought the old days, the old scenes back again, and made me for the moment younger by all those years!

A few seasons ago I feared the tribe of bluebirds were on the verge of extinction from the enormous number of them that perished from cold and hunger in the South in the winter of '94. For two summers not a blue wing, not a blue warble. I seemed to miss something kindred and precious from my environment— the visible embodiment of the tender sky and the wistful soil. What a loss, I said, to the coming generations of dwellers in the country—no bluebird in the spring ! What will the farm-boy date from ? But the fear was groundless: the birds are regaining their lost ground; broods of young blue-coats are again seen drifting from stake to stake or from mullen-stalk to mullen-stalk about the fields in summer, and our April air will doubtless again be warmed and thrilled by this lovely harbinger of spring.

JOHN BURROUGHS.

August 17, '97.

PREFACE

Not to have so much as a bowing acquaintance with the birds that nest in our gardens or under the very eaves of our houses; that haunt our wood-piles; keep our fruit-trees free from slugs; waken us with their songs, and enliven our walks along the roadside and through the woods, seems to be, at least, a breach of etiquette toward some of our most kindly disposed neighbors.

Birds of prey, game and water birds are not included in the book. The following pages are intended to be nothing more than a familiar introduction to the birds that live near us. Even in the principal park of a great city like New York, a bird-lover has found more than one hundred and thirty species; as many, probably, as could be discovered in the same sized territory anywhere.

The plan of the book is not a scientific one, if the term scientific is understood to mean technical and anatomical. The purpose of the writer is to give, in a popular and accessible form, knowledge which is accurate and reliable about the life of our common birds. This knowledge has not been collected from the stuffed carcasses of birds in museums, but gleaned afield. In a word, these short narrative descriptions treat of the bird's characteristics of size, color, and flight; its peculiarities of instinct and temperament; its nest and home life; its choice of food; its songs; and of the season in which we may expect it to play its part in the great panorama Nature unfolds with faithful precision year after year. They are an attempt to make the bird so live before the reader that, when seen out of doors, its recognition shall be instant and cordial, like that given to a friend.

The coloring described in this book is sometimes more vivid than that found in the works of some learned authorities, whose conflicting testimony is often sadly bewildering to the novice. In different parts of the country, and at different seasons of the year, the plumage of some birds undergoes many changes. The reader must remember, therefore, that the specimens examined and described were not, as before stated, the faded ones in our museums, but live birds in their fresh, spring plumage, studied afield.

The birds have been classed into color groups in the belief that this method, more than any other, will make identification most easy. The color of the bird is the first, and often the only, characteristic noticed. But they have also been classified according to the localities for which they show decided preferences and in which they are most likely to be found. Again, they have been grouped according to the season when they may be expected. In the brief paragraphs that deal with groups of birds separated into the various families represented in the book, the characteristics and traits of each clan are clearly emphasized. By these several aids it is believed the merest novice will be able to quickly identify any bird neighbor that is neither local nor rare.

To the uninitiated or uninterested observer, all small, dull-colored birds are "common sparrows." The closer scrutiny of the trained eye quickly differentiates, and picks out not only the Song, the Canada, and the Fox Sparrows, but finds a dozen other familiar friends where one who "has eyes and sees not" does not even suspect their presence. Ruskin says: "The more I think of it, I find this conclusion more impressed upon me, that the greatest thing a human soul ever does in this world is to *see* something. . . . Hundreds of people can talk for one who can think, but thousands can think for one who can see. To see clearly is poetry, prophecy, and religion—all in one."

While the author is indebted to all the time-honored standard authorities, and to many ornithologists of the present day,—too many for individual mention,—it is to Mr. John Burroughs her deepest debt is due. To this clear-visioned prophet, who has opened the blind eyes of thousands to the delights that Nature holds within our easy reach, she would gratefully acknowledge many obligations: first of all, for the plan on which "Bird Neighbors" is arranged; next, for his patient kindness in reading and annotating the manuscript of the book; and, not least, for the inspiration of his perennially charming writings that are so largely responsible for the ready-made audience now awaiting writers on out-of-door topics.

NELTJE BLANCHAN.

LIST OF COLORED PLATES

I

BIRD FAMILIES

THEIR CHARACTERISTICS AND THE
REPRESENTATIVES OF EACH FAMILY
INCLUDED IN "BIRD NEIGHBORS"

BIRD FAMILIES

THEIR CHARACTERISTICS AND THE REPRESENTATIVES OF EACH
FAMILY INCLUDED IN "BIRD NEIGHBORS"

Order Coccyges: CUCKOOS AND KINGFISHERS

Family Cuculidæ: CUCKOOS

Long, pigeon-shaped birds, whose backs are grayish brown with a bronze lustre and whose under parts are whitish. Bill long and curved. Tail long ; raised and drooped slowly while the bird is perching. Two toes point forward and two backward. Call-note loud and like a tree-toad's rattle. Song lacking. Birds of low trees and undergrowth, where they also nest ; partial to neighborhood of streams, or wherever the tent caterpillar is abundant. Habits rather solitary, silent, and eccentric. Migratory.

> Yellow-billed Cuckoo.
> Black-billed Cuckoo.

Family Alcedinidæ: KINGFISHERS

Large, top-heavy birds of streams and ponds. Usually seen perching over the water looking for fish. Head crested ; upper parts slate-blue ; underneath white, and belted with blue or rusty. Bill large and heavy. Middle and outer toes joined for half their length. Call-note loud and prolonged, like a policeman's rattle. Solitary birds ; little inclined to rove from a chosen locality. Migratory.

> Belted Kingfisher.

Order Pici: WOODPECKERS

Family Picidæ: WOODPECKERS

Medium-sized and small birds, usually with plumage black and white, and always with some red feathers about the head.

3

(The flicker is brownish and yellow instead of black and white.) Stocky, high-shouldered build ; bill strong and long for drilling holes in bark of trees. Tail feathers pointed and stiffened to serve as a prop. Two toes before and two behind for clinging. Usually seen clinging erect on tree-trunks ; rarely, if ever, head downward, like the nuthatches, titmice, etc. Woodpeckers feed as they creep around the trunks and branches. Habits rather phlegmatic. The flicker has better developed vocal powers than other birds of this class, whose rolling tattoo, beaten with their bills against the tree-trunks, must answer for their love-song. Nest in hollowed-out trees.

> Red-headed Woodpecker.
> Hairy Woodpecker.
> Downy Woodpecker.
> Yellow-bellied Woodpecker.
> Flicker.

Order Macrochires: GOATSUCKERS, SWIFTS, AND HUM-MING-BIRDS

Family Caprimulgidæ: NIGHTHAWKS, WHIPPOORWILLS, ETC.

Medium-sized, mottled brownish, gray, black, and white birds of heavy build. Short, thick head ; gaping, large mouth ; very small bill, with bristles at base. Take insect food on the wing. Feet small and weak ; wings long and powerful. These birds rest lengthwise on their perch while sleeping through the brightest daylight hours, or on the ground, where they nest.

> Nighthawk.
> Whippoorwill.

Family Micropolidæ: SWIFTS

Sooty, dusky birds seen on the wing, never resting except in chimneys of houses, or hollow trees, where they nest. Tips of tail feathers with sharp spines, used as props. They show their kinship with the goatsuckers in their nocturnal as well as diurnal habits, their small bills and large mouths for catching insects on

4

the wing, and their weak feet. Gregarious, especially at the nesting season.

Chimney Swift.

Family *Trochilidæ* : HUMMING-BIRDS

Very small birds with green plumage (iridescent red or orange breast in males); long, needle-shaped bill for extracting insects and nectar from deep-cupped flowers, and exceedingly rapid, darting flight. Small feet.

Ruby-throated Humming-bird.

Order *Passeres* : PERCHING BIRDS

Family *Tyrannidæ* : FLYCATCHERS

Small and medium-sized dull, dark-olive, or gray birds, with big heads that are sometimes crested. Bills hooked at end, and with bristles at base. Harsh or plaintive voices. Wings longer than tail ; both wings and tails usually drooped and vibrating when the birds are perching. Habits moody and silent when perching on a conspicuous limb, telegraph wire, dead tree, or fence rail and waiting for insects to fly within range. Sudden, nervous, spasmodic sallies in midair to seize insects on the wing. Usually they return to their identical perch or lookout. Pugnacious and fearless. Excellent nest builders and devoted mates.

Kingbird.
Phœbe.
Wood Pewee.
Acadian Flycatcher.
Great Crested Flycatcher.
Least Flycatcher.
Olive-sided Flycatcher.
Yellow-bellied Flycatcher.
Say's Flycatcher.

Family *Alaudidæ* : LARKS

The only true larks to be found in this country are the two species given below. They are the kin of the European skylark, of which several unsuccessful attempts to introduce the bird have been made in this country. These two larks must not be confused with the meadow larks and titlarks, which belong to the blackbird and pipit families respectively. The horned larks are birds of the ground, and are seen in the United States only in the autumn and winter. In the nesting season at the North their voices are most musical. Plumage grayish and brown, in color harmony with their habitats. Usually found in flocks ; the first species on or near the shore.

> Horned Lark.
> Prairie Horned Lark.

Family Corvidæ : CROWS AND JAYS

The crows are large black birds, walkers, with stout feet adapted for the purpose. Fond of shifting their residence at different seasons rather than strictly migratory, for, except at the northern limit of range, they remain resident all the year. Gregarious. Sexes alike. Omnivorous feeders, being partly carnivorous, as are also the jays. Both crows and jays inhabit wooded country. Their voices are harsh and clamorous ; and their habits are boisterous and bold, particularly the jays. Devoted mates ; unpleasant neighbors.

> Common Crow.
> Fish Crow.
> Northern Raven.
> Blue Jay.
> Canada Jay.

Family Icteridæ : BLACKBIRDS, ORIOLES, ETC.

Plumage black or a brilliant color combined with black. (The meadow lark a sole exception.) Sexes unlike. These birds form a connecting link between the crows and the finches. The blackbirds have strong feet for use upon the ground, where they generally feed, while the orioles are birds of the trees. They are both seed and insect eaters. The bills of the bobolink and cowbird are short and conical, for they are conspicuous seed eaters. Bills of the others long and conical, adapted for insectivorous diet. About half the family are gifted songsters.

> Red-winged Blackbird.
> Rusty Blackbird.

6

Purple Grackle.
Bronzed Grackle.
Cowbird.
Meadow Lark.
Western Meadow Lark.
Bobolink.
Orchard Oriole.
Baltimore Oriole.

Family Fringillidæ : FINCHES, SPARROWS, GROSBEAKS, BUNTINGS, LINNETS, AND CROSSBILLS

Generally fine songsters. Bills conical, short, and stout for cracking seeds. Length from five to nine inches, usually under eight inches. This, the largest family of birds that we have (about one-seventh of all our birds belong to it), comprises birds of such varied plumage and habit that, while certain family resemblances may be traced throughout, it is almost impossible to characterize the family as such. The *sparrows* are comparatively small gray and brown birds with striped upper parts, lighter underneath. Birds of the ground, or not far from it, elevated perches being chosen for rest and song. Nest in low bushes or on the ground. (Chipping sparrow often selects tall trees.) Coloring adapted to grassy, dusty habitats. Males and females similar. Flight labored. About forty species of sparrows are found in the United States ; of these, fourteen may be met with by a novice, and six, at least, surely will be.

The *finches* and their larger kin are chiefly bright-plumaged birds, the females either duller or distinct from males ; bills heavy, dull, and conical, befitting seed eaters. Not so migratory as insectivorous birds nor so restless. Mostly phlegmatic in temperament. Fine songsters.

Chipping Sparrow.
English Sparrow.
Field Sparrow.
Fox Sparrow.
Grasshopper Sparrow.
Savanna Sparrow.
Seaside Sparrow.
Sharp-tailed Sparrow.

7

Song Sparrow.
Swamp Song Sparrow.
Tree Sparrow.
Vesper Sparrow.
White-crowned Sparrow.
White-throated Sparrow.
Lapland Longspur.
Smith's Painted Longspur.
Pine Siskin (or Finch).
Purple Finch.
Goldfinch.
Redpoll.
Greater Redpoll.
Red Crossbill.
White-winged Red Crossbill.
Cardinal Grosbeak.
Rose-breasted Grosbeak.
Pine Grosbeak.
Evening Grosbeak.
Blue Grosbeak.
Indigo Bunting.
Junco.
Snowflake.
Chewink.

Family Tanagridæ : TANAGERS

Distinctly an American family, remarkable for their brilliant plumage, which, however, undergoes great changes twice a year. Females different from males, being dull and inconspicuous. Birds of the tropics, two species only finding their way north, and the summer tanager rarely found north of Pennsylvania. Shy inhabitants of woods. Though they may nest low in trees, they choose high perches when singing or feeding upon flowers, fruits, and insects. As a family, the tanagers have weak, squeaky voices, but both our species are good songsters. Suffering the fate of most bright-plumaged birds, immense numbers have been shot annually.

Scarlet Tanager.
Summer Tanager.

Family Hirundinidæ : SWALLOWS

Birds of the air, that take their insect food on the wing. Migratory. Flight strong, skimming, darting ; exceedingly graceful. When not flying they choose slender, conspicuous perches like telegraph wires, gutters, and eaves of barns. Plumage of some species dull, of others iridescent blues and greens above, whitish or ruddy below. Sexes similar. Bills small ; mouths large. Long and pointed wings, generally reaching the tip of the tail or beyond. Tail more or less forked. Feet small and weak from disuse. Song a twittering warble without power. Gregarious birds.

> Barn Swallow.
> Bank Swallow.
> Cliff (or Eaves) Swallow.
> Tree Swallow.
> Bough-winged Swallow.
> Purple Martin.

Family Ampelidæ : WAXWINGS

Medium-sized Quaker-like birds, with plumage of soft browns and grays. Head crested ; black band across forehead and through the eye. Bodies plump from indolence. Tail tipped with yellow ; wings with red tips to coverts, resembling sealing-wax. Sexes similar. Silent, gentle, courteous, elegant birds. Usually seen in large flocks feeding upon berries in the trees or perching on the branches, except at the nesting season. Voices resemble a soft, lisping twitter.

> Cedar Bird.
> Bohemian Waxwing.

Family Laniidæ : SHRIKES

Medium-sized grayish, black-and-white birds, with hooked and hawk-like bill for tearing the flesh of smaller birds, field-mice, and large insects that they impale on thorns. Handsome, bold birds, the terror of all small, feathered neighbors, not excluding the English sparrow. They choose conspicuous perches when on the lookout for prey : a projecting or dead limb of a

9

tree, the cupola of a house, the ridge-pole or weather-vane of a barn, or a telegraph wire, from which to suddenly drop upon a victim. Eyesight remarkable. Call-notes harsh and unmusical. Habits solitary and wandering. The first-named species is resident during the colder months of the year; the latter is a summer resident only north of Maryland.

Northern Shrike.
Loggerhead Shrike.

Family Vireonidæ : VIREOS OR GREENLETS

Small greenish-gray or olive birds, whitish or yellowish underneath, their plumage resembling the foliage of the trees they hunt, nest, and live among. Sexes alike. More deliberate in habit than the restless, flitting warblers that are chiefly seen darting about the ends of twigs. Vireos are more painstaking gleaners ; they carefully explore the bark, turn their heads upward to investigate the under side of leaves, and usually keep well hidden among the foliage. Bill hooked at tip for holding worms and insects. Gifted songsters, superior to the warblers. This family is peculiar to America.

Red-eyed Vireo.
Solitary Vireo.
Warbling Vireo.
White-eyed Vireo.
Yellow-throated Vireo.

Family Mniotiltidæ : WOOD WARBLERS

A large group of birds, for the most part smaller than the English sparrow ; all, except the ground warblers, of beautiful plumage, in which yellow, olive, slate-blue, black, and white are predominant colors. Females generally duller than males. Exceedingly active, graceful, restless feeders among the terminal twigs of trees and shrubbery ; haunters of tree-tops in the woods at nesting time. Abundant birds, especially during May and September, when the majority are migrating to and from regions north of the United States; but they are strangely unknown to all but devoted bird lovers, who seek them out during these months that particularly favor acquaintance. Several species are erratic in their migrations and choose a different course to return southward from the one they travelled over in spring. A few species are summer residents, and one, at least, of this tropical family, the myrtle warbler, winters at the north. The habits of the family are not identical in every representative ; some are more deliberate and less nervous than others ; a few, like the Canadian and Wilson's warblers, are expert flycatchers, taking their food on the wing, but not usually returning to the same perch, like true flycatchers ; and a few of the warblers, as, for example, the black-and-white, the pine, and the worm-eating species, have the nuthatches' habit of creeping around the bark of trees. Quite a number feed upon the ground. All are insectivorous, though many vary their diet with blossom, fruit, or berries, and naturally their bills are slender and sharply pointed, rarely finch-like. The yellow-breasted chat has the greatest variety of vocal expressions. The ground warblers are compensated for their sober, thrush-like plumage by their exquisite voices, while the great majority of the family that are gaily dressed have notes that either resemble the trill of midsummer insects or, by their limited range and feeble utterance, sadly belie the family name.

> Bay-breasted Warbler.
> Blackburnian Warbler.
> Blackpoll Warbler.
> Black-throated Blue Warbler.
> Black-throated Green Warbler.
> Black-and-white Creeping Warbler.
> Blue-winged Warbler.
> Canadian Warbler.
> Chestnut-sided Warbler.
> Golden-winged Warbler.
> Hooded Warbler.
> Kentucky Warbler.
> Magnolia Warbler.
> Mourning Warbler.
> Myrtle Warbler.
> Nashville Warbler.
> Palm Warbler.
> Parula Warbler.
> Pine Warbler.
> Prairie Warbler.

Redstart.
Wilson's Warbler.
Worm-eating Warbler.
Yellow Warbler.
Yellow Palm Warbler.
Ovenbird.
Northern Water Thrush.
Louisiana Water Thrush.
Maryland Yellowthroat.
Yellow-breasted Chat.

Family *Motacillidæ:* WAGTAILS AND PIPITS

Only three birds of this family inhabit North America, and of these only one is common enough, east of the Mississippi, to be included in this book. Terrestrial birds of open tracts near the coast, stubble-fields, and country roadsides, with brownish plumage to harmonize with their surroundings. The American pipit, or titlark, has a peculiar wavering flight when, after being flushed, it reluctantly leaves the ground. Then its white tail feathers are conspicuous. Its habit of wagging its tail when perching is not an exclusive family trait, as the family name might imply.

American Pipit, or Titlark.

Family *Troglodytidæ:* THRASHERS, WRENS, ETC.

Subfamily *Miminæ:* THRASHERS, MOCKING-BIRDS, AND CATBIRDS

Apparently the birds that comprise this large general family are too unlike to be related, but the missing links or intermediate species may all be found far South. The first subfamily is comprised of distinctively American birds. Most numerous in the tropics. Their long tails serve a double purpose—in assisting their flight and acting as an outlet for their vivacity. Usually they inhabit scrubby undergrowth bordering woods. They rank among our finest songsters, with ventriloquial and imitative powers added to sweetness of tone.

Brown Thrasher.
Catbird.
Mocking-bird.

12

AMERICAN MOCKING BIRD.
⅓ Life-size.

Subfamily *Troglodytinæ:* WRENS

Small brown birds, more or less barred with darkest brown above, much lighter below. Usually carry their short tails erect. Wings are small, for short flight. Vivacious, busy, excitable, easily displeased, quick to take alarm. Most of the species have scolding notes in addition to their lyrical, gushing song, that seems much too powerful a performance for a diminutive bird. As a rule, wrens haunt thickets or marshes, but at least one species is thoroughly domesticated. All are insectivorous.

> Carolina Wren.
> House Wren.
> Winter Wren.
> Long-billed Marsh Wren.
> Short-billed Marsh Wren.

Family *Certhiidæ:* CREEPERS

Only one species of this Old World family is found in America. It is a brown, much mottled bird, that creeps spirally around and around the trunks of trees in fall and winter, pecking at the larvæ in the bark with its long, sharp bill, and doing its work with faithful exactness but little spirit. It uses its tail as a prop in climbing, like the woodpeckers.

> Brown Creeper.

Family *Paridæ:* NUTHATCHES AND TITMICE

Two distinct subfamilies are included under this general head.

The nuthatches *(Sittinæ)* are small, slate-colored birds, seen chiefly in winter walking up and down the barks of trees, and sometimes running along the under side of branches upside down, like flies. Plumage compact and smooth. Their name is derived from their habit of wedging nuts (usually beechnuts) in the bark of trees, and then hatching them open with their strong straight bills.

> White-breasted Nuthatch.
> Red-breasted Nuthatch.

The titmice or chickadees *(Parinæ)* are fluffy little gray birds, the one crested, the other with a black cap. They are also

13

expert climbers, though not such wonderful gymnasts as the nuthatches. These cousins are frequently seen together in winter woods or in the evergreens about houses. Chickadees are partial to tree-tops, especially to the highest pine cones, on which they hang fearlessly. Cheerful, constant residents, retreating to the deep woods only to nest.

Tufted Titmouse.
Chickadee.

Family Sylviidæ : KINGLETS AND GNATCATCHERS

The kinglets (Regulinæ) are very small greenish-gray birds, with highly colored crown patch, that are seen chiefly in autumn, winter, and spring south of Labrador. Habits active ; diligent flitters among trees and shrubbery from limb to limb after minute insects. Beautiful nest builders. Song remarkable for so small a bird.

Golden-crowned Kinglet.
Ruby-crowned Kinglet.

The one representative of the distinctly American subfamily of gnatcatchers (Polioptilinæ) that we have, is a small blue-gray bird, whitish below. It is rarely found outside moist, low tracts of woodland, where insects abound. These it takes on the wing with wonderful dexterity. It is exceedingly graceful and assumes many charming postures. A bird of trees, nesting in the high branches. A bird of strong character and an exquisitely finished though feeble songster.

Blue-gray Gnatcatcher.

Family Turdidæ : THRUSHES, BLUEBIRDS, ETC.

This group includes our finest songsters. Birds of moderate size, stout build ; as a rule, inhabitants of woodlands, but the robin and the bluebird are notable exceptions. Bills long and slender, suitable for worm diet. Only casual fruit-eaters. Slender, strong legs for running and hopping. True thrushes are grayish or olive-brown above; buff or whitish below, heavily streaked or spotted.

Bluebird.
Robin.

Alice's Thrush.
Hermit Thrush.
Olive-backed Thrush.
Wilson's Thrush (Veery).
Wood Thrush.

Order Columbæ : PIGEONS AND DOVES

Family Columbidæ : PIGEONS AND DOVES

The wild pigeon is now too rare to be included among our bird neighbors ; but its beautiful relative, without the fatally gregarious habit, still nests and sings *a-coo-oo-oo* to its devoted mate in unfrequented corners of the farm or the borders of woodland. Delicately shaded fawn-colored and bluish plumage. Small heads, protruding breasts. Often seen on ground. Flight strong and rapid, owing to long wings.
Mourning or Carolina Dove.

II

HABITATS OF BIRDS

HABITATS OF BIRDS

BIRDS OF THE AIR CATCHING THEIR FOOD AS THEY FLY

Acadian Flycatcher, Great Crested Flycatcher, Least Flycatcher, Olive-sided Flycatcher, Say's Flycatcher, Yellow-bellied Flycatcher, Kingbird, Phœbe, Wood Pewee, Purple Martin, Chimney Swift, Barn Swallow, Bank Swallow, Cliff Swallow, Tree Swallow, Rough-winged Swallow, Canadian Warbler, Blackpoll, Wilson's Warbler, Nighthawk, Whippoorwill, Ruby-throated Humming-bird, Blue-gray Gnatcatcher.

BIRDS MOST FREQUENTLY SEEN IN THE UPPER HALF OF TREES

Scarlet Tanager, Summer Tanager, Baltimore Oriole, Orchard Oriole, Chickadee, Tufted Titmouse, Blue-gray Gnatcatcher, nearly all the Warblers except the Ground Warblers; Cedar Bird, Bohemian Waxwing, the Vireos, Robin, Red Crossbill, White-winged Crossbill, Purple Grackle, Bronzed Grackle, Redstart, Northern Shrike, Loggerhead Shrike, Crow, Fish Crow, Raven, Purple Finch, Tree and Chipping Sparrows, Cardinal, Blue Jay, Kingbird, the Crested and other Flycatchers.

BIRDS OF LOW TREES OR LOWER PARTS OF TREES

Black-billed Cuckoo, Yellow-billed Cuckoo, the Sparrows, the Thrushes, the Grosbeaks, Goldfinch, Summer Yellowbird and other Warblers; the Wrens, Bluebird, Mocking-bird, Catbird, Brown Thrasher, Maryland Yellowthroat, Yellow-breasted Chat.

BIRDS OF TREE-TRUNKS AND LARGE LIMBS

Hairy Woodpecker, Downy Woodpecker, Red-headed Woodpecker, Yellow-bellied Woodpecker, Flicker, White-breasted Nuthatch, Red-breasted Nuthatch, Brown Creeper, Chickadee, Tufted Titmouse, Golden-crowned Kinglet, Ruby-crowned Kinglet, Black-and-white Creeping Warbler, Blue-winged Warbler, Worm-eating Warbler, Pine Warbler, Blackpoll Warbler, Whippoorwill, Nighthawk.

BIRDS THAT SHOW A PREFERENCE FOR PINES AND OTHER EVERGREENS

Chickadee, Tufted Titmouse, the Nuthatches, Brown Creeper, the Kinglets, Pine Warbler, Black-and-white Creeping Warbler and all the Warblers except the Ground Warblers; Pine Siskin, Cedar Bird and Bohemian Waxwing (in juniper and cedar trees), Pine Grosbeak, Red Crossbill, White-winged Cross-bill, the Grackles, Crow, Raven, Pine Finch.

BIRDS SEEN FEEDING AMONG THE FOLIAGE AND TER-MINAL TWIGS OF TREES

The Red-eyed Vireo, White-eyed Vireo, Warbling Vireo, Solitary Vireo, Yellow-throated Vireo, Golden-crowned King-let, Ruby-crowned Kinglet, Black-billed Cuckoo, Yellow-billed Cuckoo, Yellow Warbler or Summer Yellowbird, nearly all the Warblers except the Pine and the Ground Warblers; the Fly-catchers, Blue-gray Gnatcatcher.

BIRDS THAT CHOOSE CONSPICUOUS PERCHES

Northern Shrike, Loggerhead Shrike, Kingbird, the Wood Pewee, the Phœbe and other Flycatchers, the Swallows, King-fisher, Crows, Grackles, Blue Jay and Canada Jay; the Song, the White-throated, and the Fox Sparrows; the Grosbeaks, Cedar Bird, Goldfinch, Robin, Purple Finch, Cowbird, Brown Thrasher while in song.

BIRDS OF THE GARDENS AND ORCHARDS

Bluebird, Robin; the English, Song, White-throated, Vesper, White-crowned, Fox, Chipping, and Tree Sparrows; Phœbe, Wood Pewee, the Least Flycatcher, Crested Flycatcher, Kingbird, Brown Thrasher, Wood Thrush, Mocking-bird, Catbird, House

Wren; nearly all the Warblers, especially at blossom time among the shrubbery and fruit trees; Cedar Bird, Purple Martin, Eaves Swallow, Barn Swallow, Purple Finch, Cowbird, Baltimore and Orchard Orioles, Purple Grackle, Bronzed Grackle, Blue Jay, Crow, Fish Crow, Chimney Swift, Ruby-throated Humming-bird, the Woodpeckers, Flicker, the Nuthatches, Chickadee, Tufted Titmouse, the Cuckoos, Mourning Dove, Junco.

BIRDS OF THE WOODS

The Warblers almost without exception; the Thrushes, the Woodpeckers, the Flycatchers, the Winter and the Carolina Wrens, the Tanagers, the Nuthatches and Titmice, the Kinglets, the Water Thrushes, the Vireos, Whippoorwill, Nighthawk, Kingfisher, Cardinal, Ovenbird, Brown Creeper, Tree Sparrow, Fox Sparrow, White-throated Sparrow, White-crowned Sparrow, Junco.

BIRDS SEEN NEAR THE EDGES OF WOODS

The Wrens, the Woodpeckers, the Flycatchers, the Warblers, Purple Finch, the Cuckoos, Brown Thrasher, Wood Thrush, Cowbird, Brown Creepers, the Nuthatches and Titmice, the Kinglets, Chewink; the White-crowned, White-throated, Tree, Fox, and Song Sparrows; Humming-bird, Bluebird, Junco, the Crossbills, the Grosbeaks, Nighthawk, Whippoorwill, Mourning Dove, Indigo Bird, Brown Thrasher.

BIRDS OF SHRUBBERY, BUSHES, AND THICKETS

Maryland Yellowthroat, Ovenbird (in woods); Myrtle Warbler, Mourning Warbler, Yellow-breasted Chat, and other Warblers during the migrations; the Shrikes; the White-throated, the Fox, the Song, and other Sparrows; Chickadee, Junco, Chewink, Rose-breasted Grosbeak, Cowbird, Red-winged Black-bird, Catbird, Mocking-bird, Wilson's Thrush, Goldfinch, Redpolls, Maryland Yellowthroat, White-eyed Vireo, Hooded Warbler.

BIRDS SEEN FEEDING ON THE GROUND

The Sparrows, Junco, Meadowlark, Horned Lark, Chewink, Robin, Ovenbird, Pipit or Titlark, Redpoll, Greater Redpoll,

Snowflake, Lapland Longspur, Smith's Painted Longspur, Rusty Blackbird, Red-winged Blackbird, the Crows, Cowbird, the Water Thrushes, Bobolink, Canada Jay, the Grackles, Mourning Dove; the Worm-eating, the Prairie, the Kentucky, and the Mourning Ground Warblers; Flicker.

BIRDS OF MEADOW, FIELD, AND UPLAND

The Field and Vesper Sparrows, Bobolink, Meadowlark, Horned Lark, Goldfinch, the Swallows, Pipit or Titlark, Cowbird, Redpoll, Greater Redpoll, Snowflake, Junco, Lapland Longspur, Smith's Painted Longspur, Rusty Blackbird, Crow, Fish Crow, Nighthawk, Whippoorwill; the Yellow, the Palm, and the Prairie Warblers; the Grackles, Flicker, Bluebird, Indigo Bird.

BIRDS OF ROADSIDE AND FENCES

The Sparrows, Kingbird, Crested Flycatcher, Yellow-breasted Chat, Indigo Bird, Bluebird, Flicker, Goldfinch, Brown Thrasher, Catbird, Robin, the Woodpeckers, Yellow Palm Warbler, the Vireos.

BIRDS OF MARSHES AND BOGGY MEADOWS

Long-billed Marsh Wren, Short-billed Marsh Wren; the Swamp, the Savanna, the Sharp-tailed, and the Seaside Sparrows; Red-winged Blackbird.

BIRDS OF WET WOODLANDS AND MARSHY THICKETS

Northern Water Thrush, Louisiana Water Thrush, Oven-bird, Winter Wren, Carolina Wren, Phœbe; Wood Pewee and the other Flycatchers; Wilson's Thrush or Veery, Blue-gray Gnatcatcher, Yellow-breasted Chat; the Canadian, Wilson's Black-capped, the Maryland Yellowthroat, the Hooded, and the Yellow-throated Warblers.

BIRDS FOUND NEAR SALT WATER

Fish Crow, Common Crow, Bank Swallow, Tree Swallow, Savanna Sparrow, Sharp-tailed Sparrow, Seaside Sparrow, Horned Lark, Pipit or Titlark.

BIRDS FOUND NEAR STREAMS AND PONDS

Kingfisher, the Swallows, Northern Water Thrush, Louisiana Water Thrush, Phœbe, Wood Pewee, the Flycatchers, Winter Wren, Wilson's Black-capped Warbler, the Canadian and the Yellow Warblers.

BIRDS THAT SING ON THE WING

Bobolink, Meadowlark, Indigo Bird, Purple Finch, Goldfinch, Ovenbird, Kingbird, Vesper Sparrow (rarely), Maryland Yellowthroat, Horned Lark, Kingfisher, the Swallows, Chimney Swift, Nighthawk, Song Sparrow, Red-winged Blackbird, Pipit or Titlark.

III

SEASONS OF BIRDS

THE LATITUDE OF NEW YORK IS TAKEN AS
AN ARBITRARY DIVISION FOR WHICH ALLOW-
ANCES MUST BE MADE FOR OTHER LOCALITIES

THE SEASONS OF BIRDS IN THE VICINITY OF NEW YORK OR, APPROXIMATELY, OF THE FORTY-SECOND DEGREE OF LATITUDE

PERMANENT RESIDENTS

Hairy Woodpecker.
Downy Woodpecker.
Yellow-bellied Woodpecker.
Red-headed Woodpecker.
Flicker.
Meadowlark.
Prairie Horned Lark.
Blue Jay.
Crow.
Fish Crow.
English Sparrow.
Social Sparrow.

Swamp Sparrow.
Song Sparrow.
Cedar Bird.
Cardinal.
Carolina Wren.
White-breasted Nuthatch.
Tufted Titmouse.
Chickadee.
Robin.
Bluebird.
Goldfinch.

WINTER RESIDENTS AND VISITORS

BIRDS SEEN BETWEEN NOVEMBER AND APRIL

English Sparrow.
Tree Sparrow.
White-throated Sparrow.
Swamp Sparrow.
Vesper Sparrow.
White-crowned Sparrow.
Fox Sparrow.
Song Sparrow.
Snowflake.
Junco.
Horned Lark.
Meadowlark.

Pine Grosbeak.
Redpoll.
Greater Redpoll.
Cedar Bird.
Bohemian Waxwing.
Hairy Woodpecker.
Downy Woodpecker.
Yellow-bellied Woodpecker.
Flicker.
Myrtle Warbler.
Northern Shrike.
White-breasted Nuthatch.

27

Red-breasted Nuthatch.
Tufted Titmouse.
Chickadee.
Robin.
Bluebird.
Ruby-crowned Kinglet.
Golden-crowned Kinglet.
Brown Creeper.
Carolina Wren.
Winter Wren.
Pipit.
Purple Finch.

Goldfinch.
Pine Siskin.
Lapland Longspur.
Smith's Painted Longspur.
Evening Grosbeak.
Cardinal.
Blue Jay.
Red Crossbill.
White-winged Crossbill.
Crow.
Fish Crow.
Kingfisher.

SUMMER RESIDENTS

BIRDS SEEN BETWEEN APRIL AND NOVEMBER

Mourning Dove.
Black-billed Cuckoo.
Yellow-billed Cuckoo.
Kingfisher.
Red-headed Woodpecker.
Hairy Woodpecker.
Downy Woodpecker.
Yellow-bellied Woodpecker.
Flicker.
Whippoorwill.
Nighthawk.
Chimney Swift.
Ruby-throated Humming-bird.
Kingbird.
Wood Pewee.
Phœbe.
Acadian Flycatcher.
Crested Flycatcher.
Least Flycatcher.
Olive-sided Flycatcher.
Yellow-bellied Flycatcher.
Say's Flycatcher.
Bobolink.
Cowbird.

Red-winged Blackbird.
Rusty Blackbird.
Orchard Oriole.
Baltimore Oriole.
Purple Grackle.
Bronzed Grackle.
Crow.
Fish Crow.
Raven.
Blue Jay.
Canada Jay.
Chipping Sparrow.
English Sparrow.
Field Sparrow.
Fox Sparrow.
Grasshopper Sparrow.
Savanna Sparrow.
Seaside Sparrow.
Sharp-tailed Sparrow.
Swamp Song Sparrow.
Song Sparrow.
Vesper Sparrow.
Rose-breasted Grosbeak.
Blue Grosbeak.

Indigo Bird.

Scarlet Tanager.

Purple Martin.

Barn Swallow.

Bank Swallow.

Cliff Swallow.

Tree Swallow.

Rough-winged Swallow.

Red-eyed Vireo.

White-eyed Vireo.

Solitary Vireo.

Warbling Vireo.

Yellow-throated Vireo.

Black-and-white Warbler.

Black-throated Green Warbler.

Blue-winged Warbler.

Chestnut-sided Warbler.

Golden-winged Warbler.

Hooded Warbler.

Pine Warbler.

Prairie Warbler.

Parula Warbler.

Worm-eating Warbler.

Yellow Warbler.

Redstart.

Ovenbird.

Northern Water Thrush.

Louisiana Water Thrush.

Yellow-breasted Chat.

Maryland Yellowthroat.

Mocking-bird.

Catbird.

Brown Thrasher.

House Wren.

Carolina Wren.

Long-billed Marsh Wren.

Short-billed Marsh Wren.

Alice's Thrush.

Hermit Thrush.

Olive-backed Thrush.

Wilson's Thrush or Veery.

Wood Thrush.

Meadowlark.

Western Meadowlark.

Prairie Horned Lark.

White-breasted Nuthatch.

Chickadee.

Tufted Titmouse.

Chewink.

Purple Finch.

Goldfinch.

Cardinal.

Robin.

Bluebird.

Cedar Bird.

Loggerhead Shrike.

SPRING AND AUTUMN MIGRANTS ONLY, OR RARE SUMMER VISITORS

The following Warblers :

Bay-breasted.

Blackburnian.

Black-polled.

Black-throated Blue.

Canadian.

Magnolia.

Mourning.

Myrtle.

Nashville.

Wilson's Black-capped.

Palm.

Yellow Palm.

Blue-gray Gnatcatcher.

Summer Tanager.

MIGRATIONS OF BIRDS IN VICINITY OF NEW YORK

FEBRUARY 15 TO MARCH 15

Bluebird, Robin, the Grackles, Song Sparrow, Fox Sparrow, Red-winged Blackbird, Kingfisher, Flicker, Purple Finch.

MARCH 15 TO APRIL 1

Increased numbers of foregoing group; Cowbird, Meadowlark, Phœbe ; the Field, the Vesper, and the Swamp Sparrows.

APRIL 1 TO 15

The White-throated and the Chipping Sparrows, the Tree and the Barn Swallows, Rusty Blackbird, the Red-headed and the Yellow-bellied Woodpeckers, Hermit Thrush, Ruby-crowned Kinglet, Pipit ; the Pine, the Myrtle, and the Yellow Palm Warblers; Goldfinch.

APRIL 15 TO MAY 1

Increased numbers of foregoing group; Brown Thrasher ; Alice's, the Olive-backed, and the Wood Thrushes ; Chimney Swift, Whippoorwill, Chewink, the Purple Martin, and the Cliff and the Bank Swallows; Least Flycatcher ; the Black-and-white Creeping, the Parula, and the Black-throated Green Warblers ; Ovenbird, House Wren, Catbird.

MAY 1 TO 15

Increased numbers of foregoing group; Wilson's Thrush or Veery; Nighthawk, Ruby-throated Humming-bird, the Cuckoos, Crested Flycatcher, Kingbird, Wood Pewee, the Marsh Wrens, Bank Swallow, the five Vireos, the Baltimore and Orchard Orioles, Bobolink, Indigo Bird, Rose-breasted Grosbeak, Scarlet Tanager, Maryland Yellowthroat, Yellow-breasted Chat, the Water Thrushes; and the Magnolia, the Yellow, the Black-throated Blue, the Bay-breasted, the Chestnut-sided, and the Golden-winged Warblers.

MAY 15 TO JUNE 1

Increased numbers of foregoing group; Yellow-bellied Flycatcher, Mocking-bird, Summer Tanager ; and the Blackburnian, the Blackpoll, the Worm-eating, the Hooded, Wilson's Black-capped, and the Canadian Warblers.

JUNE, JULY, AUGUST

In June few species of birds are not nesting; in July they may rove about more or less with their increased families, searching for their favorite foods; August finds them moulting and moping in silence, but toward the end of the month, thoughts of returning southward set them astir again.

AUGUST 15 TO SEPTEMBER 15

Bobolink, Cliff Swallow, Scarlet Tanager, Yellow-bellied Flycatcher, Purple Martin; the Blackburnian, the Worm-eating, the Bay-breasted, the Chestnut-sided, the Hooded, the Mourning, Wilson's Black-capped, and the Canadian Warblers; Baltimore Oriole, Humming-bird.

SEPTEMBER 15 TO OCTOBER 1

Increased numbers of foregoing group ; Wilson's Thrush, Wood Thrush, Kingbird, Wood Pewee, Crested Flycatcher; the Least, the Olive-sided, and the Acadian Flycatchers; the Marsh Wrens, the Cuckoos, Whippoorwill, Rose-breasted Grosbeak, Orchard Oriole, Indigo Bird; the Warbling, the Solitary, and the Yellow-throated Vireos; the Black-and-white Creeping, the Golden-winged, the Yellow, and the Black-throated Blue Warblers; Maryland Yellowthroat, Yellow-breasted Chat, Redstart.

OCTOBER 1 TO 15

Increased numbers of foregoing group; Hermit Thrush, Catbird, House Wren, Ovenbird, the Water Thrushes, the Red-eyed and the White-eyed Vireos, Wood Pewee, Nighthawk, Chimney Swift, Cowbird, Horned Lark, Winter Wren, Junco; the Tree, the Vesper, the White-throated, and the Grasshopper Sparrows; the Blackpoll, the Parula, the Pine, the Yellow Palm, and the Prairie Warblers; Chickadee, Tufted Titmouse.

31

Seasons of Birds

OCTOBER 15 TO NOVEMBER 15

Increased numbers of foregoing group; Wood Thrush, Wilson's Thrush or Veery, Alice's Thrush, Olive-backed Thrush, Robin, Chewink, Brown Thrasher, Phœbe, Shrike; the Fox, the Field, the Swamp, the Savanna, the White-crowned, the Chipping, and the Song Sparrows; the Red-winged and the Rusty Blackbirds; Meadowlark, the Grackles, Flicker, the Red-headed and the Yellow-bellied Woodpeckers; Purple Finch, the Kinglets, the Nuthatches, Pine Siskin.

IV

BIRDS GROUPED ACCORDING TO SIZE

BIRDS GROUPED ACCORDING TO SIZE

SMALLER THAN THE ENGLISH SPARROW

Humming-bird.
The Kinglets.
The Wrens.
All the Warblers not mentioned elsewhere.
Redstart.
Ovenbird.
Chickadee.
Tufted Titmouse.
Red-breasted Nuthatch.
White-breasted Nuthatch.
Blue-gray Gnatcatcher.
Yellow-bellied Flycatcher.
Acadian Flycatcher.
Least Flycatcher.

The Redpolls.
Goldfinch.
Pine Siskin.
Savanna Sparrow.
Grasshopper Sparrow.
Sharp-tailed Sparrow.
Chipping Sparrow.
Field Sparrow.
Swamp Song Sparrow.
Indigo Bunting.
Warbling Vireo.
Yellow-throated Vireo.
Red-eyed Vireo.
White-eyed Vireo.
Brown Creeper.

ABOUT THE SIZE OF THE ENGLISH SPARROW

Purple Finch.
The Crossbills.
The Longspurs.
Vesper Sparrow.
Seaside Sparrow.
Tree Sparrow.

Junco.
Song Sparrow.
Solitary Vireo.
The Water-thrushes.
Pipit or Titlark.
Downy Woodpecker.

LARGER THAN THE ENGLISH SPARROW AND SMALLER THAN THE ROBIN

Yellow-bellied Woodpecker.
Chimney Swift (apparently).
The Swallows (apparently).

Kingbird.
Crested Flycatcher.
Phœbe.

Olive-sided Flycatcher.
Wood Pewee.
Horned Lark.
Bobolink.
Cowbird.
Orchard Oriole.
Baltimore Oriole.
The Grosbeaks : Evening, Blue, Pine, Rose-breasted, and Cardinal.

Snowflake.
White-crowned Sparrow.
White-throated Sparrow.
Fox Sparrow.
The Tanagers.
Cedar Bird.
Bohemian Waxwing.
Yellow-breasted Chat.
The Thrushes.
Bluebird.

ABOUT THE LENGTH OF THE ROBIN

Red-headed Woodpecker.
Hairy Woodpecker.
Red-winged Blackbird.
Rusty Blackbird.
Loggerhead Shrike.

Northern Shrike.
Mocking-bird.
Catbird.
Chewink.
Purple Martin (apparently).

LONGER THAN THE ROBIN

Mourning Dove.
The Cuckoos.
Kingfisher.
Flicker.
Raven.
Crow.
Fish Crow.

Blue Jay.
Canada Jay.
Meadowlark.
Whippoorwill (apparently)
Nighthawk (apparently)
The Grackles.
Brown Thrasher.

36

V

DESCRIPTIONS OF BIRDS

GROUPED ACCORDING TO COLOR

BIRDS CONSPICUOUSLY BLACK

Common Crow
Fish Crow
American Raven
Purple Grackle
Bronzed Grackle
Rusty Blackbird
Red-winged Blackbird
Purple Martin
Cowbird

See also several of the Swallows; the Kingbird, the Phœbe, the Wood Pewee, and other Flycatchers; the Chimney Swift; and the Chewink.

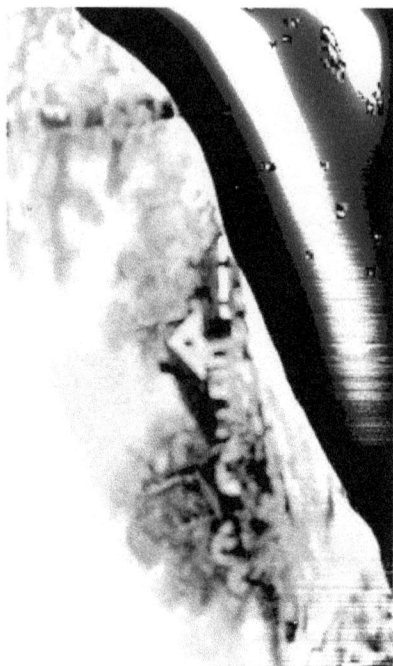

BIRDS CONSPICUOUSLY BLACK

The Common Crow

(Corvus Americanus) Crow family

Called also : CORN THIEF

Length—16 to 17.50 inches.
Male—Glossy black with violet reflections. Wings appear saw-
   toothed when spread, and almost equal the tail in length.
Female—Like male, except that the black is less brilliant.
Range—Throughout North America, from Hudson Bay to the
   Gulf of Mexico.
Migrations—March. October. Summer and winter resident.

If we have an eye for the picturesque, we place a certain
value upon the broad, strong dash of color in the landscape, given
by a flock of crows flapping their course above a corn-field, against
an October sky ; but the practical eye of the farmer looks only
for his gun in such a case. To him the crow is an unmitigated
nuisance, all the more maddening because it is clever enough to
circumvent every means devised for its ruin. Nothing escapes
its rapacity ; fear is unknown to it. It migrates in broad daylight, chooses the most conspicuous perches, and yet its assur-
ance is amply justified in its steadily increasing numbers.

In the very early spring, note well the friendly way in which
the crow follows the plow, ingratiating itself by eating the larvæ,
field mice, and worms upturned in the furrows, for this is its one
serviceable act throughout the year. When the first brood of
chickens is hatched, its serious depredations begin. Not only
the farmer's young fledglings, ducks, turkeys, and chicks, are
snatched up and devoured, but the nests of song birds are made
desolate, eggs being crushed and eaten on the spot, when there
are no birds to carry off to the rickety, coarse nest in the high
tree top in the woods. The fish crow, however, is the much

greater enemy of the birds. Like the common crows, this, their smaller cousin, likes to congregate in winter along the seacoast to feed upon shell-fish and other sea-food that the tide brings to its feet.

Samuels claims to have seen a pair of crows visit an orchard and destroy the young in two robins' nests in half an hour. He calculates that two crows kill, in one day alone, young birds that in the course of the season would have eaten a hundred thousand insects. When, in addition to these atrocities, we remember the crow's depredations in the corn-field, it is small wonder that among the first laws enacted in New York State was one offering a reward for its head. But the more scientific agriculturists now concede that the crow is the farmer's true friend.

Fish Crow

(*Corvus ossifragus*) Crow family

Length—14 to 16 inches. About half as large again as the robin.
Male and Female—Glossy black, with purplish-blue reflections, generally greener underneath. Chin naked.
Range—Along Atlantic coast and that of the Gulf of Mexico, northward to southern New England. Rare stragglers on the Pacific coast.
Migrations—March or April. September. Summer resident only at northern limit of range. Is found in Hudson River valley about half-way to Albany.

Compared with the common crow, with which it is often confounded, the fish crow is of much smaller, more slender build. Thus its flight is less labored and more like a gull's, whose habit of catching fish that may be swimming near the surface of the water it sometimes adopts. Both Audubon and Wilson, who first made this species known, record its habit of snatching food as it flies over the southern waters—a rare practice at the north. Its plumage, too, differs slightly from the common crow's in being a richer black everywhere, and particularly underneath, where the "corn thief" is dull. But it is the difference between the two crows' call-note that we chiefly depend upon to distinguish these confusing cousins. To say that the fish crow says *car-r-r* instead of a loud, clear *caw*, means little until we have had an opportunity to compare its hoarse, cracked voice with the other bird's familiar call.

From the farmer's point of view, there is still another distinction: the fish crow lets his crops alone. It contents itself with picking up refuse on the shores of the sea or rivers not far inland; haunting the neighborhood of fishermen's huts for the small fish discarded when the seines are drawn, and treading out with its toes the shell-fish hidden in the sand at low tide. When we see it in the fields it is usually intent upon catching field-mice, grubs, and worms, with which it often varies its fish diet. It is, however, the worst nest robber we have ; it probably destroys ten times as many eggs and young birds as its larger cousin.

The fishermen have a tradition that this southern crow comes and goes with the shad and herring—a saw which science unkindly disapproves.

American Raven

(Corvus corax principalis) Crow family

Called also : NORTHERN RAVEN

Length—26 to 27 inches. Nearly three times as large as a robin.
Male and Female—Glossy black above, with purplish and greenish reflections. Duller underneath. Feathers of the throat and breast long and loose, like fringe.
Range—North America, from polar regions to Mexico. Rare along Atlantic coast and in the south. Common in the west, and very abundant in the northwest.
Migrations—An erratic wanderer, usually resident where it finds its way.

The weird, uncanny voice of this great bird that soars in wide circles above the evergreen trees of dark northern forests seems to come out of the skies like the malediction of an evil spirit. Without uttering the words of any language—Poe's "Nevermore" was, of course, a poetic license—people of all nationalities appear to understand that some dire calamity, some wicked portent, is being announced every time the unbirdlike creature utters its rasping call. The superstitious folk crow with an "I told you so," as they solemnly wag their heads when they hear of some death in the village after "the bird of ill-omen" has made an unwelcome visit to the neighborhood. It receives the blame for every possible misfortune.

When seen in the air, the crow is the only other bird for which the raven could be mistaken; but the raven does more sailing and less flapping, and he delights in describing circles as he easily soars high above the trees. On the ground, he is seen to be a far larger bird than the largest crow. The curious beard or fringe of feathers on his breast at once distinguishes him.

These birds show the family instinct for living in flocks large and small, not of ravens only, but of any birds of their own genera. In the art of nest building they could instruct most of their relatives. High up in evergreen trees or on the top of cliffs, never very near the seashore, they make a compact, symmetrical nest of sticks, neatly lined with grasses and wool from the sheep pastures, adding soft, comfortable linings to the old nest from year to year for each new brood. When the young emerge from the eggs, which take many curious freaks of color and markings, they are pied black and white, suggesting the young of the western white-necked raven, a similarity which, so far as plumage is concerned, they quickly outgrow. They early acquire the fortunate habit of eating whatever their parents set before them—grubs, worms, grain, field-mice; anything, in fact, for the raven is a conspicuously omnivorous bird.

Purple Grackle

(Quiscalus quiscula) Blackbird family

Called also: CROW BLACKBIRD; MAIZE THIEF; KEEL-TAILED GRACKLE

Length—12 to 13 inches. About one-fourth as large again as the robin.

Male—Iridescent black, in which metallic violet, blue, copper, and green tints predominate. The plumage of this grackle has iridescent bars. Iris of eye bright yellow and conspicuous. Tail longer than wings.

Female—Less brilliant black than male, and smaller.

Range—Gulf of Mexico to 57th parallel north latitude.

Migrations—Permanent resident in Southern States. Few are permanent throughout range. Migrates in immense flocks in March and September.

BRONZED GRACKLE
⅓ Life size.

This "refined crow" (which is really no crow at all except in appearance) has scarcely more friends than a thief is entitled to ; for, although in many sections of the country it has given up its old habit of stealing Indian corn and substituted ravages upon the grasshoppers instead, it still indulges a crow-like instinct for pillaging nests and eating young birds.

Travelling in immense flocks of its own kind, a gregarious bird of the first order, it nevertheless is not the social fellow that its cousin, the red-winged blackbird, is. It especially holds aloof from mankind, and mankind reciprocates its suspicion.

The tallest, densest evergreens are not too remote for it to build its home, according to Dr. Abbott, though in other States than New Jersey, where he observed them, an old orchard often contains dozens of nests. One peculiarity of the grackles is that their eggs vary so much in coloring and markings that different sets examined in the same groups of trees are often wholly unlike. The average groundwork, however, is soiled blue or greenish, waved, streaked, or clouded with brown. These are laid in a nest made of miscellaneous sticks and grasses, rather carefully constructed, and lined with mud. Another peculiarity is the bird's method of steering itself by its tail when it wishes to turn its direction or alight.

Peering at you from the top of a dark pine tree with its staring yellow eye, the grackle is certainly uncanny. There, very early in the spring, you may hear its cracked and wheezy whistle, for, being aware that however much it may look like a crow it belongs to another family, it makes a ridiculous attempt to sing. When a number of grackles lift up their voices at once, some one has aptly likened the result to a "good wheel-barrow chorus !" The grackle's mate alone appreciates his efforts as, standing on tiptoe, with half-spread wings and tail, he pours forth his craven soul to her through a disjointed larynx.

With all their faults, and they are numerous, let it be recorded of both crows and grackles that they are as devoted lovers as turtle-doves. Lowell characterizes them in these four lines :

> " Fust come the black birds, clatt'rin' in tall trees,
> And settlin' things in windy Congresses ;
> Queer politicians, though, for I'll be skinned
> If all on 'em don't head against the wind."

. . . . .

45

The *Bronzed Grackle* (*Quiscalus quiscula æneus*) differs from the preceding chiefly in the more brownish bronze tint of its plumage and its lack of iridescent bars. Its range is more westerly, and in the southwest it is particularly common ; but as a summer resident it finds its way to New England in large numbers. The call-note is louder and more metallic than the purple grackle's. In nearly all respects the habits of these two birds are identical.

Rusty Blackbird

(Scolecophagus carolinus) Blackbird family

Called also : THRUSH BLACKBIRD ; RUSTY GRACKLE ; RUSTY ORIOLE ; RUSTY CROW ; BLACKBIRD

Length—9 to 9.55 inches. A trifle smaller than the robin.

Male—In full plumage, glossy black with metallic reflections, intermixed with rusty brown that becomes more pronounced as the season advances. Pale straw-colored eyes.

Female—Duller plumage and more rusty, inclining to gray. Light line over eye. Smaller than male.

Range—North America, from Newfoundland to Gulf of Mexico and westward to the Plains.

Migrations—April. November. A few winter north.

A more sociable bird than the grackle, though it travel in smaller flocks, the rusty blackbird condescends to mingle freely with other feathered friends in marshes and by brooksides. You can identify it by its rusty feathers and pale yellow eye, and easily distinguish the rusty-gray female from the female redwing that is conspicuously streaked.

In April flocks of these birds may frequently be seen along sluggish, secluded streams in the woods, feeding upon the seeds of various water or brookside plants, and probably upon insects also. At such times they often indulge in a curious spluttering, squeaking, musical concert that one listens to with pleasure. The breeding range is mostly north of the United States. But little seems to be known of the birds' habits in their northern home.

Why it should ever have been called a thrush blackbird is one of those inscrutable mysteries peculiar to the naming of birds

46

which are so frequently called precisely what they are not. In spite of the compliment implied in associating the name of one of our finest songsters with it, the rusty blackbird has a clucking call as unmusical as it is infrequent, and only very rarely in the spring does it pipe a note that even suggests the sweetness of the redwing's.

Red-winged Blackbird

(*Agelaius phœniceus*) Blackbird family

Called also : SWAMP BLACKBIRD ; RED-WINGED ORIOLE ; RED-WINGED STARLING

Length—Exceptionally variable—7.50 to 9.80 inches. Usually about an inch smaller than the robin.
Male—Coal-black. Shoulders scarlet, edged with yellow.
Female—Feathers finely and inconspicuously speckled with brown, rusty black, whitish, and orange. Upper wing-coverts rusty black, tipped with white, or rufous and sometimes spotted with black and red.
Range—North America. Breeds from Texas to Columbia River, and throughout the United States. Commonly found from Mexico to 57th degree north latitude.
Migrations—March. October. Common summer resident.

In oozy pastures where a brook lazily finds its way through the farm is the ideal pleasure ground of this "bird of society." His notes, "*h'-wa-ker-ee*" or "*con-quer-ee*" (on an ascending scale), are liquid in quality, suggesting the sweet, moist, cool retreats where he nests. Liking either heat or cold (he is fond of wintering in Florida, but often retreats to the north while the marshes are still frozen); enjoying not only the company of large flocks of his own kind with whom he travels, but any bird associates with whom he can scrape acquaintance ; or to sit quietly on a tree-top in the secluded, inaccessible bog while his mate is nesting; satisfied with cut-worms, grubs, and insects, or with fruit and grain for his food—the blackbird is an impressive and helpful example of how to get the best out of life.

Yet, of all the birds, some farmers complain that the blackbird is the greatest nuisance. They dislike the noisy chatterings when a flock is simply indulging its social instincts. They

47

complain, too, that the blackbirds eat their corn, forgetting that having devoured innumerable grubs from it during the summer, the birds feel justly entitled to a share of the profits. Though occasionally guilty of eating the farmer's corn and oats and rice, yet it has been found that nearly seven-eighths of the red-wing's food is made up of weed-seeds or of insects injurious to agriculture.

This bird builds its nest in low bushes on the margin of ponds or low in the bog grass of marshes. From three to five pale-blue eggs, curiously streaked, spotted, and scrawled with black or purple, constitute a brood. Nursery duties are soon finished, for in July the young birds are ready to gather in flocks with their elders.

> " The blackbirds make the maples ring
> With social cheer and jubilee ;
> The red-wing flutes his ' O-ka-lee ! ' "
>
> —*Emerson.*

Purple Martin

(*Progne subis*) Swallow family

Length—7 to 8 inches. Two or three inches smaller than the robin.

Male—Rich glossy black with bluish and purple reflections ; duller black on wings and tail. Wings rather longer than the tail, which is forked.

Female—More brownish and mottled ; grayish below.

Range—Peculiar to America. Penetrates from Arctic Circle to South America.

Migrations—Late April. Early September. Summer resident.

In old-fashioned gardens, set on a pole over which honey-suckle and roses climbed from a bed where China pinks, phlox, sweet Williams, and hollyhocks crowded each other below, martin boxes used always to be seen with a pair of these large, beautiful swallows circling overhead. But now, alas! the boxes, where set up at all, are quickly monopolized by the English sparrow, a bird that the martin, courageous as a kingbird in attacking crows and hawks, tolerates as a neighbor only when it must.

Bradford Torrey tells of seeing quantities of long-necked squashes dangling from poles about the negro cabins all through

RED-WINGED BLACK BIRD.

the South. One day he asked an old colored man what these
squashes were for.

"Why, deh is martins' boxes," said Uncle Remus. "No
danger of hawks carryin' off de chickens so long as de martins
am around."

The Indians, too, have always had a special liking for this
bird. They often lined a hollowed-out gourd with bits of bark
and fastened it in the crotch of their tent poles to invite its friendship. The Mohegan Indians have called it "the bird that never
rests"—a name better suited to the tireless barn swallow, Dr.
Abbott thinks.

Wasps, beetles, and all manner of injurious garden insects
constitute its diet—another reason for its universal popularity.
It is simple enough to distinguish the martins from the other
swallows by their larger size and iridescent dark coat, not to
mention their song, which is very soft and sweet, like musical
laughter, rippling up through the throat.

Cowbird

(Molothrus ater) Blackbird family

Called also : BROWN-HEADED ORIOLE ; COW-PEN BIRD ;
COW BLACKBIRD ; COW BUNTING

Length—7 to 8 inches. About one-fifth smaller than the robin.

Male—Iridescent black, with head, neck, and breast glistening
brown. Bill dark brown, feet brownish.

Female—Dull grayish-brown above, a shade lighter below, and
streaked with paler shades of brown.

Range—United States, from coast to coast. North into British
America, south into Mexico.

Migrations—March. November. Common summer resident.

The cowbird takes its name from its habit of walking about
among the cattle in the pasture, picking up the small insects
which the cattle disturb in their grazing. The bird may often
be seen within a foot or two of the nose of a cow or heifer, walking briskly about like a miniature hen, intently watching for its
insect prey.

Its marital and domestic character is thoroughly bad.

49

Conspicuously Black

Polygamous and utterly irresponsible for its offspring, this bird forms a striking contrast to other feathered neighbors, and indeed is almost an anomaly in the animal kingdom. In the breeding season an unnatural mother may be seen skulking about in the trees and shrubbery, seeking for nests in which to place a surreptitious egg, never imposing it upon a bird of its size, but selecting in a cowardly way a small nest, as that of the vireos or warblers or chipping sparrows, and there leaving the hatching and care of its young to the tender mercies of some already burdened little mother. It has been seen to remove an egg from the nest of the red-eyed vireo in order to place one of its own in its place. Not finding a convenient nest, it will even drop its eggs on the ground, trusting them to merciless fate, or, still worse, devouring them. The eggs are nearly an inch long, white speckled with brown or gray.

Cowbirds are gregarious. The ungrateful young birds, as soon as they are able to go roaming, leave their foster-parents and join the flock of their own kind. In keeping with its unclean habits and unholy life and character, the cowbird's ordinary note is a gurgling, rasping whistle, followed by a few sharp notes.

BIRDS CONSPICUOUSLY BLACK AND WHITE

Red-headed Woodpecker

Hairy Woodpecker

Downy Woodpecker

Yellow-bellied Woodpecker

Chewink

Snowflake

Rose-breasted Grosbeak

Bobolink

Black-poll Warbler

Black-and-white Creeping Warbler

See also the Swallows; the Shrikes; Nuthatches and Titmice; the Kingbird and other Flycatchers; the Nighthawk; the Redstart; and the following Warblers: the Myrtle; the Bay-breasted; the Blackburnian; and the Black-throated Blue Warbler.

BIRDS CONSPICUOUSLY BLACK AND WHITE

Red-headed Woodpecker

(Melanerpes erythrocephalus) Woodpecker family

Called also : TRI-COLOR ; RED-HEAD

Length—8.50 to 9.75 inches. An inch or less smaller than the robin.

Male and Female—Head, neck, and throat crimson; breast and underneath white; back black and white; wings and tail blue black, with broad white band on wings conspicuous in flight.

Range—United States, east of Rocky Mountains and north to Manitoba.

Migrations—Abundant but irregular migrant. Most commonly seen in Autumn, and rarely resident.

In thinly populated sections, where there are few guns about, this is still one of the commonest as it is perhaps the most conspicuous member of the woodpecker family, but its striking glossy black-and-white body and its still more striking crimson head, flattened out against the side of a tree like a target, where it is feeding, have made it all too tempting a mark for the rifles of the sportsmen and the sling-shots of small boys. As if sufficient attention were not attracted to it by its plumage, it must needs keep up a noisy, guttural rattle, *ker-r-ruck, ker-r-ruck,* very like a tree-toad's call, and flit about among the trees with the restlessness of a fly-catcher. Yet, in spite of these invitations for a shot to the passing gunner, it still multiplies in districts where nuts abound, being "more common than the robin" about Washington, says John Burroughs.

All the familiar woodpeckers have two characteristics most prominently exemplified in this red-headed member of their tribe. The hairy, the downy, the crested, the red-bellied, the sapsucker, and the flicker have each a red mark somewhere about

53

their heads as if they had been wounded there and bled a little—
some more, some less ; and the figures of all of them, from much
flattening against tree-trunks, have become high-shouldered and
long-waisted.

The red-headed woodpecker selects, by preference, a partly
decayed tree in which to excavate a hole for its nest, because
the digging is easier, and the sawdust and chips make a softer
lining than green wood. Both male and female take turns in
this hollowing-out process. The one that is off duty is allowed
"twenty minutes for refreshments," consisting of grubs, beetles,
ripe apples or cherries, corn, or preferably beech-nuts. At a
loving call from its mate in the hollow tree, it returns promptly
to perform its share of the work, when the carefully observed
"time is up." The heap of sawdust at the bottom of the hollow
will eventually cradle from four to six glossy-white eggs.

This woodpecker has the thrifty habit of storing away nuts
in the knot-holes of trees, between cracks in the bark, or in
decayed fence rails—too often a convenient storehouse at which
the squirrels may help themselves. But it is the black snake that
enters the nest and eats the young family, and that is a more
deadly foe than even the sportsman or the milliner.

The Hairy Woodpecker

(Dryobates villosus) Woodpecker family

Length—9 to 10 inches. About the size of the robin.
Male—Black and white above, white beneath. White stripe
down the back, composed of long hair-like feathers. Bright-
red band on the nape of neck. Wings striped and dashed
with black and white. Outer tail feathers white, without
bars. White stripe about eyes and on sides of the head.
Female—Without the red band on head, and body more brownish than that of the male.
Range—Eastern parts of United States, from the Canadian border to the Carolinas.
Migrations—Resident throughout its range.

The bill of the woodpecker is a hammering tool, well fitted
for its work. Its mission in life is to rid the trees of insects,

RED-HEADED WOODPECKER.
Life-size.

which hide beneath the bark, and with this end in view, the bird is seen clinging to the trunks and branches of trees through fair and wintry weather, industriously scanning every inch for the well-known signs of the boring worm or destructive fly.

In the autumn the male begins to excavate his winter quarters, carrying or throwing out the chips, by which this good workman is known, with his beak, while the female may make herself cosey or not, as she chooses, in an abandoned hole. About her comfort he seems shamefully unconcerned. Intent only on his own, he drills a perfectly round hole, usually on the under side of a limb where neither snow nor wind can harm him, and digs out a horizontal tunnel in the dry, brittle wood in the very heart of the tree, before turning downward into the deep, pear-shaped chamber, where he lives in selfish solitude. But when the nesting season comes, how devoted he is temporarily to the mate he has neglected and even abused through the winter ! Will she never learn that after her clear-white eggs are laid and her brood raised he will relapse into the savage and forget all his tender wiles ?

The hairy woodpecker, like many another bird and beast, furnishes much doubtful weather lore for credulous and inexact observers. "When the woodpecker pecks low on the trees, expect warm weather" is a common saying, but when different individuals are seen pecking at the same time, one but a few feet from the ground, and another among the high branches, one may make the prophecy that pleases him best.

The hairy woodpeckers love the deep woods. They are drummers, not singers, but when walking in the desolate winter woods even the drumming and tapping of the busy feathered workmen on a resonant limb is a solace, giving a sense of life and cheerful activity which is invigorating.

The Downy Woodpecker

(Dryobates pubescens) Woodpecker family

Length—6 to 7 inches. About the size of the English sparrow.
Male—Black above, striped with white. Tail shaped like a wedge. Outer tail feathers white, and barred with black. Middle tail feathers black. A black stripe on top of head, and distinct white band over and under the eyes. Red patch on upper

side of neck. Wings, with six white bands crossing them transversely; white underneath.

Female—Similar, but without scarlet on the nape, which is white.

Range—Eastern North America, from Labrador to Florida.

Migrations—Resident all the year throughout its range.

The downy woodpecker is similar to his big relative, the hairy woodpecker, in color and shape, though much smaller. His outer tail feathers are white, barred with black, but the hairy's white outer tail feathers lack these distinguishing marks.

He is often called a sapsucker—though quite another bird alone merits that name—from the supposition that he bores into the trees for the purpose of sucking the sap ; but his tongue is ill adapted for such use, being barbed at the end, and most ornithologists consider the charge libellous. It has been surmised that he bores the numerous little round holes close together, so often seen, with the idea of attracting insects to the luscious sap. The woodpeckers never drill for insects in live wood. The downy actually drills these little holes in apple and other trees to feed upon the inner milky bark of the tree—the cambium layer. The only harm to be laid to his account is that, in his zeal, he sometimes makes a ring of small holes so continuous as to inadvertently damage the tree by girdling it. The bird, like most others, does not debar himself entirely from fruit diet, but enjoys berries, especially poke-berries.

He is very social with birds and men alike. In winter he attaches himself to strolling bands of nuthatches and chickadees, and in summer is fond of making friendly visits among village folk, frequenting the shade trees of the streets and grapevines of back gardens. He has even been known to fearlessly peck at flies on window panes.

In contrast to his large brother woodpecker, who is seldom drawn from timber lands, the little downy member of the family brings the comfort of his cheery presence to country homes, beating his rolling tattoo in spring on some resonant limb under our windows in the garden with a strength worthy of a larger drummer.

This rolling tattoo, or drumming, answers several purposes: by it he determines whether the tree is green or hollow; it startles insects from their lurking places underneath the bark, and it also serves as a love song.

YELLOW-BELLIED SAPSUCKER.
$\frac{2}{3}$ Life size.

Yellow-bellied Woodpecker

(Sphyrapicus varius) Woodpecker family

Called also : THE SAPSUCKER

Length—8 to 8.6 inches. About one-fifth smaller than the robin.
Male—Black, white, and yellowish white above, with bright-red
  crown, chin, and throat.   Breast black, in form of crescent.
  A yellowish-white line, beginning at bill and passing below
  eye, merges into the pale yellow of the bird underneath.
  Wings spotted with white, and coverts chiefly white.   Tail
  black; white on middle of feathers.
Female—Paler, and with head and throat white.
Range—Eastern North America, from Labrador to Central America.
Migrations—April.   October.   Resident north of Massachusetts.
  Most common in autumn.

It is sad to record that this exquisitely marked woodpecker,
the most jovial and boisterous of its family, is one of the very
few bird visitors whose intimacy should be discouraged.   For its
useful appetite for slugs and insects which it can take on the
wing with wonderful dexterity, it need not be wholly con-
demned.   But as we look upon a favorite maple or fruit tree
devitalized or perhaps wholly dead from its ravages, we cannot
forget that this bird, while a most abstemious fruit-eater, has a
pernicious and most intemperate thirst for sap.   Indeed, it spends
much of its time in the orchard, drilling holes into the freshest,
most vigorous trees ; then, when their sap begins to flow, it
siphons it into an insatiable throat, stopping in its orgie only
long enough to snap at the insects that have been attracted to
the wounded tree by the streams of its heart-blood now trickling
down its sides.   Another favorite pastime is to strip the bark off
a tree, then peck at the soft wood underneath—almost as fatal a
habit.   It drills holes in maples in early spring for sap only.   If it
drills holes in fruit trees it is for the cambium layer, a soft, pulpy,
nutritious under-bark.

These woodpeckers have a variety of call-notes, but their
rapid drumming against the limbs and trunks of trees is the
sound we always associate with them and the sound that Mr.
Bicknell says is the love-note of the family.

Unhappily, these birds, that many would be glad to have

decrease in numbers, take extra precautions for the safety of their young by making very deep excavations for their nests, often as deep as eighteen or twenty inches.

The Chewink

(*Pipilo erythrophthalmus*) Finch family

Called also : GROUND ROBIN ; TOWHEE ; TOWHEE BUNT-ING ; TOWHEE GROUND FINCH ; GRASEL

Length—8 to 8.5 inches. About one-fifth smaller than the robin.
Male—Upper parts black, sometimes margined with rufous. Breast white; chestnut color on sides and rump. Wings marked with white. Three outer feathers of tail striped with white, conspicuous in flight. Bill black and stout. Red eyes; feet brown.
Female—Brownish where the male is black. Abdomen shading from chestnut to white in the centre.
Range—From Labrador, on the north, to the Southern States ; west to the Rocky Mountains.
Migrations—April. September and October. Summer resident. Very rarely a winter resident at the north.

The unobtrusive little chewink is not infrequently mistaken for a robin, because of the reddish chestnut on its under parts. Careful observation, however, shows important distinctions. It is rather smaller and darker in color; its carriage and form are not those of the robin, but of the finch. The female is smaller still, and has an olive tint in her brown back. Her eggs are inconspicuous in color, dirty white speckled with brown, and laid in a sunken nest on the ground. Dead leaves and twigs abound, and form, as the anxious mother fondly hopes, a safe hiding place for her brood. So careful concealment, however, brings peril to the fledglings, for the most cautious bird-lover may, and often does, inadvertently set his foot on the hidden nest.

The chewink derives its name from the fancied resemblance of its note to these syllables, while those naming it "towhee" hear the sound *to-whick, to-whick, to-whee.* Its song is rich, full, and pleasing, and given only when the bird has risen to the branches above its low foraging ground.

It frequents the border of swampy places and bushy fields.

SNOW BUNTING.

It is generally seen in the underbrush, picking about among the dead leaves for its steady diet of earthworms and larvæ of insects, occasionally regaling itself with a few dropping berries and fruit.

When startled, the bird rises not more than ten or twelve feet from the earth, and utters its characteristic calls. On account of this habit of flying low and grubbing among the leaves, it is sometimes called the ground robin. In the South our modest and useful little food-gatherer is often called grasel, especially in Louisiana, where it is white-eyed, and is much esteemed, alas! by epicures.

Snowflake

(Plectrophenax nivalis) Finch family

Called also : SNOW BUNTING ; WHITEBIRD ; SNOWBIRD ; SNOW LARK

Length—7 to 7.5 inches. About one-fourth smaller than the robin.
Male and Female—Head, neck, and beneath soiled white, with a few reddish-brown feathers on top of head, and suggesting an imperfect collar. Above, grayish brown obsoletely streaked with black, the markings being most conspicuous in a band between shoulders. Lower tail feathers black; others, white and all edged with white. Wings brown, white, and gray. Plumage unusually variable. In summer dress (in arctic regions) the bird is almost white.
Range—Circumpolar regions to Kentucky (in winter only).
Migrations—Midwinter visitor; rarely, if ever, resident south of arctic regions.

These snowflakes (mentioned collectively, for it is impossible to think of the bird except in great flocks) are the " true spirits of the snowstorm," says Thoreau. They are animated beings that ride upon it, and have their life in it. By comparison with the climate of the arctic regions, no doubt our hardiest winter weather seems luxuriously mild to them. We associate them only with those wonderful midwinter days when sky, fields, and woods alike are white, and a " hard, dull bitterness of cold " drives every other bird and beast to shelter. It is said they often pass the night buried beneath the snow. They have been seen to dive beneath it to escape a hawk.

Whirling about in the drifting snow to catch the seeds on

59

the tallest stalks that the wind in the open meadows uncovers,
the snowflakes suggest a lot of dead leaves being blown through
the all-pervading whiteness. Beautiful soft brown, gray, and
predominating black-and-white coloring distinguish these capricious visitors from the slaty junco, the "snowbird" more commonly known. They are, indeed, the only birds we have that
are nearly white; and rarely, if ever, do they rise far above the
ground their plumage so admirably imitates.

At the far north, travellers have mentioned their inspiriting
song, but in the United States we hear only their cheerful twitter.
Nansen tells of seeing an occasional snow bunting in that desola-
tion of arctic ice where the *Fram* drifted so long.

The Rose-breasted Grosbeak

(Habia ludoviciana) Finch family

Length—7.75 to 8.5 inches. About one-fifth smaller than the robin.
Male—Head and upper parts black. Breast has rose-carmine
   shield-shaped patch, often extending downward to the centre
   of the abdomen. Underneath, tail quills, and two spots on
   wings white. Conspicuous yellow, blunt beak.
Female—Brownish, with dark streakings, like a sparrow. No
   rose-color. Light sulphur yellow under wings. Dark brown,
   heavy beak.
Range—Eastern North America, from southern Canada to Panama.
Migrations—Early May. September. Summer resident.

A certain ornithologist tells with complacent pride of having
shot over fifty-eight rose-breasted grosbeaks in less than three
weeks (during the breeding season) to learn what kind of food
they had in their crops. This kind of devotion to science may
have quite as much to do with the growing scarcity of this bird
in some localities as the demands of the milliners, who, however,
receive all of the blame for the slaughter of our beautiful songsters.
The farmers in Pennsylvania, who, with more truth than poetry,
call this the potato-bug bird, are taking active measures, however, to protect the neighbor that is more useful to their crop than
all the insecticides known. It also eats flies, wasps, and grubs.

Seen upon the ground, the dark bird is scarcely attractive with
his clumsy beak overbalancing a head that protrudes with stupid-

ROSE-BREASTED GROSSBEAKS.
⅓ Life-size.

looking awkwardness; but as he rises into the trees his lovely rose-colored breast and under-wing feathers are seen, and before he has had time to repeat his delicious, rich-voiced warble you are already in love with him. Vibrating his wings after the manner of the mocking-bird, he pours forth a marvellously sweet, clear, mellow song (with something of the quality of the oriole's, robin's, and thrush's notes), making the day on which you first hear it memorable. This is one of the few birds that sing at night. A soft, sweet, rolling warble, heard when the moon is at its full on a midsummer night, is more than likely to come from the rose-breasted grosbeak.

It is not that his quiet little sparrow-like wife has advanced notions of feminine independence that he takes his turn at sitting upon the nest, but that he is one of the most unselfish and devoted of mates. With their combined efforts they construct only a coarse, unlovely cradle in a thorn-bush or low tree near an old, overgrown pasture lot. The father may be the poorest of architects, but as he patiently sits brooding over the green, speckled eggs, his beautiful rosy breast just showing above the grassy rim, he is a sufficient adornment for any bird's home.

The Bobolink

(*Dolichonyx oryzivorus*) Blackbird family

Called also : REEDBIRD; MAYBIRD; MEADOW-BIRD; AMERI-CAN ORTOLAN; BUTTER-BIRD; SKUNK BLACKBIRD

Length—7 inches. A trifle larger than the English sparrow.

Male—In spring plumage : black, with light-yellow patch on upper neck, also on edges of wings and tail feathers. Rump and upper wings splashed with white. Middle of back streaked with pale buff. Tail feathers have pointed tips. *In autumn plumage*, resembles female.

Female—Dull yellow-brown, with light and dark dashes on back, wings, and tail. Two decided dark stripes on top of head.

Range—North America, from eastern coast to western prairies. Migrates in early autumn to Southern States, and in winter to South America and West Indies.

Migrations—Early May. From August to October. Common summer resident.

Conspicuously Black and White

Perhaps none of our birds have so fitted into song and story as the bobolink. Unlike a good child, who should "be seen and not heard," he is heard more frequently than seen. Very shy, of peering eyes, he keeps well out of sight in the meadow grass before entrancing our listening ears. The bobolink never soars like the lark, as the poets would have us believe, but generally sings on the wing, flying with a peculiar self-conscious flight horizontally thirty or forty feet above the meadow grass. He also sings perched upon the fence or tuft of grass. He is one of the greatest *poseurs* among the birds.

In spring and early summer the bobolinks respond to every poet's effort to imitate their notes. "Dignified 'Robert of Lincoln' is telling his name," says one; "Spink, spank, spink," another hears him say. But best of all are Wilson Flagg's lines:

> . . . "Now they rise and now they fly ;
> They cross and turn, and in and out, and down the middle and wheel about,
> With a ' Phew, shew, Wadolincon ; listen to me Bobolincon ! ' "

After midsummer the cares of the family have so worn upon the jollity of our dashing, rollicking friend that his song is seldom heard. The colors of his coat fade into a dull yellowish brown like that of his faithful mate, who has borne the greater burden of the season, for he has two complete moults each year.

The bobolinks build their nest on the ground in high grass. The eggs are of a bluish white. Their food is largely insectivorous: grasshoppers, crickets, beetles, spiders, with seeds of grass especially for variety.

In August they begin their journey southward, flying mainly by night. Arriving in the Southern States, they become the sad-colored, low-voiced rice or reed bird, feeding on the rice fields, where they descend to the ignominious fate of being dressed for the plate of the epicure.

Could there be a more tragic ending to the glorious note of the gay songster of the north?

BOBOLINKS.
⅔ Life-size.

Blackpoll Warbler

(*Dendroica striata*) Wood Warbler family

Length—5.5 to 6 inches. About an inch smaller than the English sparrow.

Male—Black cap; cheeks and beneath grayish white, forming a sort of collar, more or less distinct. Upper parts striped gray, black, and olive. Breast and under parts white, with black streaks. Tail olive-brown, with yellow-white spots.

Female—Without cap. Greenish-olive above, faintly streaked with black. Paler than male. Bands on wings, yellowish.

Range—North America, to Greenland and Alaska. In winter, to northern part of South America.

Migrations—Last of May. Late October.

A faint "*screep, screep,*" like "the noise made by striking two pebbles together," Audubon says, is often the only indication of the blackpoll's presence; but surely that tireless bird-student had heard its more characteristic notes, which, rapidly uttered, increasing in the middle of the strain and diminishing toward the end, suggest the shrill, wiry hum of some midsummer insect. After the opera-glass has searched him out we find him by no means an inconspicuous bird. A dainty little fellow, with a glossy black cap pulled over his eyes, he is almost hidden by the dense foliage on the trees by the time he returns to us at the very end of spring. Giraud says that he is the very last of his tribe to come north, though the bay-breasted warbler has usually been thought the bird to wind up the spring procession.

The blackpoll has a certain characteristic motion that distinguishes him from the black-and-white creeper, for which a hasty glance might mistake him, and from the jolly little chickadee with his black cap. Apparently he runs about the tree-trunk, but in reality he so flits his wings that his feet do not touch the bark at all; yet so rapidly does he go that the flipping wing-motion is not observed. He is most often seen in May in the apple trees, peeping into the opening blossoms for insects, uttering now and then his slender, lisping, brief song.

Vivacious, a busy hunter, often catching insects on the wing like the flycatchers, he is a cheerful, useful neighbor the short time he spends with us before travelling to the far north, where he mates and nests. A nest has been found on Slide Mountain, in the Catskills, but the hardy evergreens of Canada, and some-

63

times those of northern New England, are the chosen home of
this little bird that builds a nest of bits of root, lichens, and sedges,
amply large for a family twice the size of his.

Black-and-white Creeping Warbler

(Mniotilta varia) Wood Warbler family

Called also : VARIED CREEPING WARBLER ; BLACK-AND-
WHITE CREEPER ; WHITEPOLL WARBLER

Length—5 to 5.5 inches. About an inch smaller than the English
sparrow.

Male—Upper parts white, varied with black. A white stripe
along the summit of the head and back of the neck, edged
with black. White line above and below the eye. Black
cheeks and throat, grayish in females and young. Breast
white in middle, with black stripes on sides. Wings and
tail rusty black, with two white cross-bars on former, and
soiled white markings on tail quills.

Female—Paler and less distinct markings throughout.

Range—Peculiar to America. Eastern United States and westward to the plains. North as far as the fur countries. Winters in tropics south of Florida.

Migrations—April. Late September. Summer resident.

Nine times out of ten this active little warbler is mistaken for
the downy woodpecker, not because of his coloring alone, but
also on account of their common habit of running up and down
the trunks of trees and on the under side of branches, looking for
insects, on which all the warblers subsist. But presently the true
warbler characteristic of restless flitting about shows itself. A
woodpecker would go over a tree with painstaking, systematic
care, while the black-and-white warbler, no less intent upon
securing its food, hurries off from tree to tree, wherever the most
promising *menu* is offered.

Clinging to the mottled bark of the tree-trunk, which he so
closely resembles, it would be difficult to find him were it not
for these sudden flittings and the feeble song, "*Weachy,
weachy, weachy, 'twee, 'twee, 'tweet,*" he half lisps, half sings
between his dashes after slugs. Very rarely indeed can his nest
be found in an old stump or mossy bank, where bark, leaves,
and hair make the downy cradle for his four or five tiny babies.

BLACK AND WHITE CREEPING WARBLER.
Life-size.

DUSKY AND GRAY AND SLATE-COLORED BIRDS

Chimney Swift
Kingbird
Wood Pewee
Phœbe and Say's Phœbe
Crested Flycatcher
Olive-sided Flycatcher
Least Flycatcher
Chickadee
Tufted Titmouse
Canada Jay
Catbird
Mocking-bird
Junco
White-breasted Nuthatch
Red-breasted Nuthatch
Loggerhead Shrike
Northern Shrike
Bohemian Waxwing
Bay-breasted Warbler
Chestnut-sided Warbler
Golden-winged Warbler
Myrtle Warbler
Parula Warbler
Black-throated Blue Warbler

See also the Grayish Green and the Grayish Brown Birds, particularly the Cedar Bird, several Swallows, the Acadian and the Yellow-bellied Flycatchers; Alice's and the Olive-backed Thrushes; the Louisiana Water Thrush; the Blue-gray Gnatcatcher; and the Seaside Sparrow. See also the females of the following birds: Pine Grosbeak; White-winged Red Crossbill; Purple Martin; and the Nashville, the Pine, and the Magnolia Warblers.

CHIMNEY SWIFT.
⅔ Life-size.

DUSKY, GRAY, AND SLATE-COLORED BIRDS

Chimney Swift

(Chætura pelagica) Swift family

Called also : CHIMNEY SWALLOW ; AMERICAN SWIFT

Length—5 to 5.45 inches. About an inch shorter than the Eng-
lish sparrow   Long wings make its length appear greater.
Male and Female—Deep sooty gray ; throat of a trifle lighter gray.
Wings extend an inch and a half beyond the even tail, which
has sharply pointed and very elastic quills, that serve as props.
Feet are muscular, and have exceedingly sharp claws.
Range—Peculiar to North America east of the Rockies, and from
Labrador to Panama.
Migrations—April.   September or October.   Common summer
resident.

The chimney swift is, properly speaking, not a swallow at
all, though chimney swallow is its more popular name. Rowing
towards the roof of your house, as if it used first one wing, then
the other, its flight, while swift and powerful, is stiff and mechan-
ical, unlike the swallow's, and its entire aspect suggests a bat.
The nighthawk and whippoorwill are its relatives, and it resembles them not a little, especially in its nocturnal habits.

So much fault has been found with the misleading names of
many birds, it is pleasant to record the fact that the name of the
chimney swift is everything it ought to be. No other birds can
surpass and few can equal it in its powerful flight, sometimes
covering a thousand miles in twenty-four hours, it is said, and
never resting except in its roosting places (hollow trees or chimneys of dwellings), where it does not perch, but rather clings to
the sides with its sharp claws, partly supported by its sharper
tail. Audubon tells of a certain plane tree in Kentucky where
he counted over nine thousand of these swifts clinging to the
hollow trunk.

Their nest, which is a loosely woven twig lattice, made of twigs of trees, which the birds snap off with their beaks and carry in their beaks, is glued with the bird's saliva or tree-gum into a solid structure, and firmly attached to the inside of chimneys, or hollow trees where there are no houses about. Two broods in a season usually emerge from the pure white, elongated eggs.

What a twittering there is in the chimney that the swifts appropriate after the winter fires have died out! Instead of the hospitable column of smoke curling from the top, a cloud of sooty birds wheels and floats above it. A sound as of distant thunder fills the chimney as a host of these birds, startled, perhaps, by some indoor noise, whirl their way upward. Woe betide the happy colony if a sudden cold snap in early summer necessitates the starting of a fire on the hearth by the unsuspecting householder! The glue being melted by the fire, "down comes the cradle, babies and all" into the glowing embers. A prolonged, heavy rain also causes their nests to loosen their hold and fall with the soot to the bottom.

Thrifty New England housekeepers claim that bedbugs, commonly found on bats, infest the bodies of swifts also, which is one reason why wire netting is stretched across the chimney tops before the birds arrive from the South.

Kingbird

(*Tyrannus tyrannus*) Flycatcher family

Called also: TYRANT FLYCATCHER; BEE MARTIN

Length—8 inches. About two inches shorter than the robin.
Male—Ashy black above; white, shaded with ash-color, beneath. A concealed crest of orange-red on crown. Tail black, terminating with a white band conspicuous in flight. Wing feathers edged with white. Feet and bill black.
Female—Similar to the male, but lacking the crown.
Range—United States to the Rocky Mountains. British provinces to Central and South America.
Migrations—May. September. Common summer resident.

If the pugnacious propensity of the kingbird is the occasion of its royal name, he cannot be said to deserve it from any fine or noble qualities he possesses. He is a born fighter from the very

WOOD PEWEE.
⅔ Life-size.

love of it, without provocation, rhyme, or reason. One can but watch with a degree of admiration his bold sallies on the big, black crow or the marauding hawk, but when he bullies the small inoffensive birds in wanton attacks for sheer amusement, the charge is less entertaining. Occasionally, when the little victim shows pluck and faces his assailant, the kingbird will literally turn tail and show the white feather. His method of attack is always when a bird is in flight; then he swoops down from the telegraph pole or high point of vantage, and strikes on the head or back of the neck, darting back like a flash to the exact spot from which he started. By these tactics he avoids a return blow and retreats from danger. He never makes a fair hand-to-hand fight, or whatever is equivalent in bird warfare. It is a satisfaction to record that he does not attempt to give battle to the catbird, but whenever in view makes a grand detour to give him a wide berth.

The kingbird feeds on beetles, canker-worms, and winged insects, with an occasional dessert of berries. He is popularly supposed to prefer the honeybee as his favorite tidbit, but the weight of opinion is adverse to the charge of his depopulating the beehive, even though he owes his appellation bee martin to this tradition. One or two ornithologists declare that he selects only the drones for his diet, which would give him credit for marvellous sight in his rapid motion through the air. The kingbird is preëminently a bird of the garden and orchard. The nest is open, though deep, and not carefully concealed. Eggs are nearly round, bluish white spotted with brown and lilac. With truly royal exclusiveness, the tyrant favors no community of interest, but sits in regal state on a conspicuous throne, and takes his grand flights alone or with his queen, but never with a flock of his kind.

Wood Pewee

(Contopus virens) Flycatcher family

Length—6.50 inches. A trifle larger than the English sparrow.

Male—Dusky brownish olive above, darkest on head ; paler on throat, lighter still underneath, and with a yellowish tinge on the dusky gray under parts. Dusky wings and tail, the wing coverts tipped with soiled white, forming two indistinct bars. Whitish eye-ring. Wings longer than tail.

Female—Similar, but slightly more buff underneath.

Dusky, Gray, and Slate-colored

Range—Eastern North America, from Florida to northern British
provinces. Winters in Central America.
Migrations—May. October. Common summer resident.

The wood pewee, like the olive-sided flycatcher, has wings
decidedly longer than its tail, and it is by no means a simple
matter for the novice to tell these birds apart or separate them
distinctly in the mind from the other members of a family whose
coloring and habits are most confusingly similar. This dusky
haunter of tall shady trees has not yet learned to be sociable like
the phœbe, but while it may not be so much in evidence close
to our homes, it is doubtless just as common. The orchard is as
near the house as it often cares to come. An old orchard, where
modern insecticides are unknown and neglect allows insects to
riot among the decayed bark and fallen fruit, is a happy hunting
ground enough, but the bird's real preferences are decidedly for
high tree-tops in the woods, where no sunshine touches the
feathers on his dusky coat. It is one of the few shade-loving
birds. In deep solitudes, where it surely retreats by nesting
time, however neighborly it may be during the migrations, its
pensive, pathetic notes, long drawn out, seem like the expression
of some hidden sorrow. *Pe-a-wee, pe-a-wee, pewee-ah-peer* is the
burden of its plaintive song, a sound as depressing as it is familiar
in every walk through the woods, and the bird's most prominent
characteristic.

To see the bird dashing about in his aërial chase for insects,
no one would accuse him of melancholia. He keeps an eye on
the "main chance," whatever his preying grief may be, and
never allows it to affect his appetite. Returning to his perch
after a successful sally in pursuit of the passing fly, he repeats his
"sweetly solemn thought" over and over again all day long and
every day throughout the summer.

The wood pewees show that devotion to each other and to
their home, characteristic of their family. Both lovers work on
the construction of the flat nest that is saddled on some mossy or
lichen-covered limb, and so cleverly do they cover the rounded
edge with bits of bark and lichen that sharp eyes only can detect
where the cradle lies. Creamy-white eggs, whose larger end is
wreathed with brown and lilac spots, are guarded with fierce
solicitude.

Trowbridge has celebrated this bird in a beautiful poem.

Phœbe

(Sayornis phœbe) Flycatcher family

Called also : DUSKY FLYCATCHER ; BRIDGE PEWEE ; WATER PEWEE

Length—7 inches. About an inch longer than the English sparrow.
Male and Female—Dusky olive-brown above ; darkest on head, which is slightly crested. Wings and tail dusky, the outer edges of some tail feathers whitish. Dingy yellowish white underneath. Bill and feet black.
Range—North America, from Newfoundland to the South Atlantic States, and westward to the Rockies. Winters south of the Carolinas, into Mexico, Central America, and the West Indies.
Migrations—March. October. Common summer resident.

The earliest representative of the flycatcher family to come out of the tropics where insect life fairly swarms and teems, what does the friendly little phœbe find to attract him to the north in March while his prospective dinners must all be still in embryo ? He looks dejected, it is true, as he sits solitary and silent on some projecting bare limb in the garden, awaiting the coming of his tardy mate ; nevertheless, the date of his return will not vary by more than a few days in a given locality year after year. Why birds that are mated for life, as these are said to be, and such devoted lovers, should not travel together on their journey north, is another of the many mysteries of bird-life awaiting solution.

The reunited, happy couple go about the garden and outbuildings like domesticated wrens, investigating the crannies on piazzas, where people may be coming and going, and boldly entering barn-lofts to find a suitable site for the nest that it must take much of both time and skill to build.

Pewit, phœbe, phœbe ; pewit, phœbe, they contentedly but rather monotonously sing as they investigate all the sites in the neighborhood. Presently a location is chosen under a beam or rafter, and the work of collecting moss and mud for the foundation and hair and feathers or wool to line the exquisite little home begins. But the labor is done cheerfully, with many a sally in midair either to let off superfluous high spirits or to catch a morsel on the wing, and with many a vivacious outburst of what by courtesy only we may name a song.

71

When not domesticated, as these birds are rapidly becoming, the phœbes dearly love a cool, wet woodland retreat. Here they hunt and bathe ; here they also build in a rocky bank or ledge of rocks or underneath a bridge, but always with clever adaptation of their nest to its surroundings, out of which it seems a natural growth. It is one of the most finished, beautiful nests ever found.

A pair of phœbes become attached to a spot where they have once nested; they never stray far from it, and return to it regularly, though they may not again occupy the old nest. This is because it soon becomes infested with lice from the hen's feathers used in lining it, for which reason too close relationship with this friendly bird-neighbor is discouraged by thrifty housekeepers. When the baby birds have come out from the four or six little white eggs, their helpless bodies are mercilessly attacked by parasites, and are often so enfeebled that half the brood die. The next season another nest will be built near the first, the following summer still another, until it would appear that a colony of birds had made their homes in the place.

Throughout the long summer—for as the phœbe is the first flycatcher to come, so it is the last to go—the bird is a tireless hunter of insects, which it catches on the wing with a sharp click of its beak, like the other members of its dexterous family.

Say's Phœbe *(Sayornis saya)* is the Western representative of the Eastern species, which it resembles in coloring and many of its habits. It is the bird of the open plains, a tireless hunter in midair sallies from an isolated perch, and has the same vibrating motion of the tail that the Eastern phœbe indulges in when excited. This bird differs chiefly in its lighter coloring, but not in habits, from the black pewee of the Pacific slope.

Great-crested Flycatcher

(Myiarchus crinitus) Flycatcher family

Called also : CRESTED FLYCATCHER

Length—8.50 to 9 inches. A little smaller than the robin.
Male and Female—Feathers of the head pointed and erect. Upper
    parts dark grayish-olive, inclining to rusty brown on wings
    and tail. Wing coverts crossed with two irregular bars of
    yellowish white. Throat gray, shading into sulphur-yellow

PHOEBE.
⅔ Life-size.

underneath, that also extends under the wings. Inner vane of several tail quills rusty red. Bristles at base of bill.

Range—From Mexico, Central America, and West Indies northward to southern Canada and westward to the plains. Most common in Mississippi basin ; common also in eastern United States, south of New England.

Migrations—May. September. Common summer resident.

The most dignified and handsomely dressed member of his family, the crested flycatcher has, nevertheless, an air of pensive melancholy about him when in repose that can be accounted for only by the pain he must feel every time he hears himself screech. His harsh, shrill call, louder and more disagreeable than the kingbird's, cannot but rasp his ears as it does ours. And yet it is chiefly by this piercing note, given with a rising inflection, that we know the bird is in our neighborhood ; for he is somewhat of a recluse, and we must often follow the disagreeable noise to its source in the tree-tops before we can catch a glimpse of the screecher. Perched on a high lookout, he appears morose and sluggish, in spite of his aristocratic-looking crest, trim figure, and feathers that must seem rather gay to one of his dusky tribe. A low soliloquy, apparently born of discontent, can be overheard from the foot of his tree. But another second, and he has dashed off in hot pursuit of an insect flying beyond our sight, and with extremely quick, dexterous evolutions in midair, he finishes the hunt with a sharp click of his bill as it closes over the unhappy victim, and then he returns to his perch. On the wing he is exceedingly active and joyous; in the tree he appears just the reverse. That he is a domineering fellow, quite as much of a tyrant as the notorious kingbird, that bears the greater burden of opprobrium, is shown in the fierce way he promptly dashes at a feathered stranger that may have alighted too near his perch, and pursues it beyond the bounds of justice, all the while screaming his rasping cry into the intruder's ears, that must pierce as deep as the thrusts from his relentless beak. He has even been known to drive off woodpeckers and bluebirds from the hollows in the trees that he, like them, chooses for a nest, and appropriate the results of their labor for his scarcely less belligerent mate. With a slight but important and indispensable addition, the stolen nest is ready to receive her four cream-colored eggs, that look as if a pen dipped in purple ink had been scratched over them.

73

The fact that gives the great-crested flycatcher a unique interest among all North American birds is that it invariably lines its nest with snake-skins if one can be had. Science would scarcely be worth the studying if it did not set our imaginations to work delving for plausible reasons for Nature's strange doings. Most of us will doubtless agree with Wilson (who made a special study of these interesting nests and never found a single one without cast snake-skins in it, even in districts where snakes were so rare they were supposed not to exist at all), that the lining was chosen to terrorize all intruders. The scientific mind that is unwilling to dismiss any detail of Nature's work as merely arbitrary and haphazard, is greatly exercised over the reason for the existence of crests on birds. But, surely, may not the sight of snake-skins that first greet the eyes of the fledgling flycatchers as they emerge from the shell be a good and sufficient reason why the feathers on their little heads should stand on end? "In the absence of a snake-skin, I have found an onion skin and shad scales in the nest," says John Burroughs, who calls this bird "the wild Irishman of the flycatchers."

Olive-sided Flycatcher

(Contopus borealis) Flycatcher family

Length—7 to 7.5 inches. About an inch longer than the English sparrow.

Male and Female—Dusky olive or grayish brown above; head darkest. Wings and tail blackish brown, the former sometimes, but not always, margined and tipped with dusky white. Throat yellowish white ; other under parts slightly lighter shade than above. Olive-gray on sides. A tuft of yellowish-white, downy feathers on flanks. Bristles at base of bill.

Range—From Labrador to Panama. Winters in the tropics. Nests usually north of United States, but it also breeds in the Catskills.

Migrations—May. September. Resident only in northern part of its range.

Only in the migrations may people south of Massachusetts hope to see this flycatcher, which can be distinguished from the rest of its kin by the darker under parts, and by the fluffy, yellowish-white tufts of feathers on its flanks. Its habits have the family characteristics: It takes its food on the wing, suddenly sallying forth from its perch, darting about midair to seize its prey, then as suddenly returning to its identical point of vantage, usually in some distended, dead limb in the tree-top; it is pugnacious, bold, and tyrannical; mopish and inert when not on the hunt, but wonderfully alert and swift when in pursuit of insect or feathered foe. The short necks of the flycatchers make their heads appear large for their bodies, a peculiarity slightly emphasized in this member of the family.

High up in some evergreen tree, well out on a branch, over which the shapeless mass of twigs and moss that serves as a nest is saddled, four or five buff-speckled eggs are laid, and by some special dispensation rarely fall out of their insecure cradle. A sharp, loud whistle, *wheu—o-wheu-o-wheu-o*, rings out from the throat of this olive-sided tyrant, warning all intruders off the premises ; but however harshly he may treat the rest of the feathered world, he has only gentle devotion to offer his brooding mate.

Least Flycatcher

(Empidonax minimus) Flycatcher family

Called also : CHEBEC

Length—5 to 5.5 inches. About an inch smaller than the English sparrow.

Male—Gray or olive-gray above, paler on wings and lower part of back, and a more distinct olive-green on head. Underneath grayish white, sometimes faintly suffused with pale yellow. Wings have whitish bars. White eye-ring. Lower half of bill horn-color.

Female—Is slightly more yellowish underneath.

Range—Eastern North America, from tropics northward to Quebec.

Migrations—May. September. Common summer resident.

This, the smallest member of its family, takes the place of the more southerly Acadian flycatcher, throughout New England and the region of the Great Lakes. But, unlike his Southern relative, he prefers orchards and gardens close to our homes for his hunting grounds rather than the wet recesses of the forests. *Che-bec, che-bec*, the diminutive olive-pated gray sprite calls out

from the orchard between his aërial sallies after the passing insects
that have been attracted by the decaying fruit, and chebec is the
name by which many New Englanders know him.

While giving this characteristic call-note, with drooping,
jerking tail, trembling wings, and uplifted parti-colored bill, he
looks unnerved and limp by the effort it has cost him. But in
the next instant a gnat flies past. How quickly the bird recovers
itself, and charges full-tilt at his passing dinner! The sharp click
of his little bill proves that he has not missed his aim; and after
careering about in the air another minute or two, looking for
more game to snap up on the wing, he will return to the same
perch and take up his familiar refrain. Without hearing this call-
note one might often mistake the bird for either the wood pewee
or the phœbe, for all the three are similarly clothed and have
many traits in common. The slightly larger size of the phœbe
and pewee is not always apparent when they are seen perching
on the trees. Unlike the "tuft of hay" to which the Acadian
flycatcher's nest has been likened, the least flycatcher's home is
a neat, substantial cup-shaped cradle softly lined with down or
horsehair, and placed generally in an upright crotch of a tree, well
above the ground.

The Chickadee

(*Parus atricapillus*) Titmouse family

Called also : BLACK-CAPPED TITMOUSE ; BLACK-CAP TIT

Length—5 to 5.5 inches. About an inch smaller than the English
   sparrow.

Male and Female—Not crested. Crown and nape and throat
   black. Above gray, slightly tinged with brown. A white
   space, beginning at base of bill, extends backwards, widen-
   ing over cheeks and upper part of breast, forming a sort of
   collar that almost surrounds neck. Underneath dirty white,
   with pale rusty-brown wash on sides. Wings and tail gray,
   with white edgings. Plumage downy.

Range—Eastern North America. North of the Carolinas to Labrador. Does not migrate in the North.

Migrations—Late September. May. Winter resident ; permanent resident in northern parts of the United States.

No "fair weather friend" is the jolly little chickadee. In the
depth of the autumn equinoctial storm it returns to the tops of

CHICKADEE.
Life-size.

the trees close by the house, where, through the sunshine, snow, and tempest of the entire winter, you may hear its cheery, irrepressible *chickadee-dee-dee-dee* or *day-day-day* as it swings around the dangling cones of the evergreens. It fairly overflows with good spirits, and is never more contagiously gay than in a snowstorm. So active, so friendly and cheering, what would the long northern winters be like without this lovable little neighbor?

It serves a more utilitarian purpose, however, than bracing faint-hearted spirits. "There is no bird that compares with it in destroying the female canker-worm moths and their eggs," writes a well-known entomologist. He calculates that as a chickadee destroys about 5,500 eggs in one day, it will eat 138,750 eggs in the twenty-five days it takes the canker-worm moth to crawl up the trees. The moral that it pays to attract chickadees about your home by feeding them in winter is obvious. Mrs. Mabel Osgood Wright, in her delightful and helpful book "Birdcraft," tells us how she makes a sort of a bird-hash of finely minced raw meat, waste canary-seed, buckwheat, and cracked oats, which she scatters in a sheltered spot for all the winter birds. The way this is consumed leaves no doubt of its popularity. A raw bone, hung from an evergreen limb, is equally appreciated.

Friendly as the chickadee is—and Dr. Abbott declares it the tamest bird we have—it prefers well-timbered districts, especially where there are red-bud trees, when it is time to nest. It is very often clever enough to leave the labor of hollowing out a nest in the tree-trunk to the woodpecker or nuthatch, whose old homes it readily appropriates ; or, when these birds object, a knot-hole or a hollow fence-rail answers every purpose. Here, in the summer woods, when family cares beset it, a plaintive, minor whistle replaces the *chickadee-dee-dee* that Thoreau likens to "silver tinkling" as he heard it on a frosty morning.

> " Piped a tiny voice near by,
> Gay and polite, a cheerful cry—
> Chick-chickadeedee ! saucy note
> Out of sound heart and merry throat,
> As if it said, ' Good-day, good Sir !
> Fine afternoon, old passenger !
> Happy to meet you in these places
> Where January brings few faces.' "
>
> —*Emerson.*

Tufted Titmouse

(*Parus bicolor*) Titmouse family

Called also CRESTED TITMOUSE; CRESTED TOMTIT

Length—6 to 6.5 inches.   About the size of the English sparrow.

Male and Female—Crest high and pointed.   Leaden or ash-gray above; darkest on wings and tail.   Frontlet, bill, and shoulders black; space between eyes gray.   Sides of head dull white.   Under parts light gray; sides yellowish, tinged with red.

Range—United States east of plains, and only rarely seen so far north as New England.

Migrations—October.   April.   Winter resident, but also found throughout the year in many States.

"A noisy titmouse is Jack Frost's trumpeter" may be one of those few weather-wise proverbs with a grain of truth in them. As the chickadee comes from the woods with the frost, so it may be noticed his cousin, the crested titmouse, is in more noisy evidence throughout the winter.

One might sometimes think his whistle, like a tugboat's, worked by steam.   But how effectually nesting cares alone can silence it in April !

Titmice always see to it you are not lonely as you walk through the woods.   This lordly tomtit, with his jaunty crest, keeps up a persistent whistle at you as he flits from tree to tree, leading you deeper into the forest, calling out " *Here-here-here!* " and looking like a pert and jaunty little blue jay, minus his gay clothes.   Mr. Nehrling translates one of the calls " *Heedle-dee-dle-dee-dle-dee!* " and another " *Peto-peto-peto-daytee-daytee!* " But it is at the former, sharply whistled as the crested titmouse gives it, that every dog pricks up his ears.

Comparatively little has been written about this bird, because it is not often found in New England, where most of the bird *litterateurs* have lived.   South of New York State, however, it is a common resident, and much respected for the good work it does in destroying injurious insects, though it is more fond of varying its diet with nuts, berries, and seeds than that all-round benefactor, the chickadee.

78

Canada Jay

(Perisoreus canadensis) Crow and Jay family

Called also: WHISKY JACK OR JOHN; MOOSE-BIRD; MEAT-
BIRD; VENISON HERON; GREASE-BIRD; CANADIAN
CARRION-BIRD; CAMP ROBBER

Length—11 to 12 inches. About two inches larger than the robin.
Male and Female—Upper parts gray; darkest on wings and tail;
back of the head and nape of the neck sooty, almost black.
Forehead, throat, and neck white, and a few white tips on
wings and tail. Underneath lighter gray. Tail long. Plumage fluffy.
Range—Northern parts of the United States and British provinces
of North America.
Migrations—Resident where found.

The Canada jay looks like an exaggerated chickadee, and
both birds are equally fond of bitter cold weather, but here the
similarity stops short. Where the chickadee is friendly the jay is
impudent and bold; hardly less of a villain than his blue relative
when it comes to marauding other birds' nests and destroying
their young. With all his vices, however, intemperance cannot
be attributed to him, in spite of the name given him by the Adirondack lumbermen and guides. "Whisky John" is a purely
innocent corruption of "Wis-ka-tjon," as the Indians call this
bird that haunts their camps and familiarly enters their wigwams.
The numerous popular names by which the Canada jays are
known are admirably accounted for by Mr. Hardy in a bulletin
issued by the Smithsonian Institution.

"They will enter the tents, and often alight on the bow of a
canoe, where the paddle at every stroke comes within eighteen
inches of them. I know nothing which can be eaten that they
will not take, and I had one steal all my candles, pulling them
out endwise, one by one, from a piece of birch bark in which
they were rolled, and another peck a large hole in a keg of cas-
tile soap. A duck which I had picked and laid down for a few
minutes, had the entire breast eaten out by one or more of these
birds. I have seen one alight in the middle of my canoe and
peck away at the carcass of a beaver I had skinned. They often
spoil deer saddles by pecking into them near the kidneys. They

do great damage to the trappers by stealing the bait from traps set for martens and minks and by eating trapped game. They will sit quietly and see you build a log trap and bait it, and then, almost before your back is turned, you hear their hateful *ca-ca-ca!* as they glide down and peer into it. They will work steadily, carrying off meat and hiding it. I have thrown out pieces, and watched one to see how much he would carry off. He flew across a wide stream, and in a short time looked as bloody as a butcher from carrying large pieces; but his patience held out longer than mine. I think one would work as long as Mark Twain's California jay did trying to fill a miner's cabin with acorns through a knot-hole in the roof. They are fond of the berries of the mountain ash, and, in fact, few things come amiss; I believe they do not possess a single good quality except industry."

One virtue not mentioned by Mr. Hardy is their prudent saving from the summer surplus to keep the winter storeroom well supplied like a squirrel's. Such thrift is the more necessary when a clamorous, hungry family of young jays must be reared while the thermometer is often as low as thirty degrees below zero at the end of March. How eggs are ever hatched at all in a temperature calculated to freeze any sitting bird stiff, is one of the mysteries of the woods. And yet four or five fluffy little jays, that look as if they were dressed in gray fur, emerge from the eggs before the spring sunshine has unbound the icy rivers or melted the snowdrifts piled high around the evergreens.

Catbird

(Galeoscoptes carolinensis) Mocking-bird family

Called also : BLACK-CAPPED THRUSH

Length—9 inches. An inch shorter than the robin.

Male and Female—Dark slate above; below somewhat paler; top of head black. Distinct chestnut patch under the tail, which is black; feet and bill black also. Wings short, more than two inches shorter than the tail.

Range—British provinces to Mexico; west to Rocky Mountains, rarely to Pacific coast. Winters in Southern States, Central America, and Cuba.

Migrations—May. November. Common summer resident.

CATBIRD.
⅔ Life-size.

Our familiar catbird, of all the feathered tribe, presents the most contrary characteristics, and is therefore held in varied estimation—loved, admired, ridiculed, abused. He is the veriest "Dr. Jekyll and Mr. Hyde" of birds. Exquisitely proportioned, with finely poised black head and satin-gray coat, which he bathes most carefully and prunes and prinks by the hour, he appears from his toilet a Beau Brummell, an aristocratic-looking, even dandified neighbor. Suddenly, as if shot, he drops head and tail and assumes the most hang-dog air, without the least sign of self-respect; then crouches and lengthens into a roll, head forward and tail straightened, till he looks like a little, short gray snake, lank and limp. Anon, with a jerk and a sprint, every muscle tense, tail erect, eyes snapping, he darts into the air intent upon some well-planned mischief. It is impossible to describe his various attitudes or moods. In song and call he presents the same opposite characteristics. How such a bird, exquisite in style, can demean himself to utter such harsh, altogether hateful catcalls and squawks as have given the bird his common name, is a wonder when in the next moment his throat swells and beginning *phut-phut-coquillicat*, he gives forth a long glorious song, only second to that of the wood thrush in melody. He is a jester, a caricaturist, a mocking-bird.

The catbird's nest is like a veritable scrap-basket, loosely woven of coarse twigs, bits of newspaper, scraps, and rags, till this rough exterior is softly lined and made fit to receive the four to six pretty dark green-blue eggs to be laid therein.

As a fruit thief harsh epithets are showered upon the friendly, confiding little creature at our doors; but surely his depredations may be pardoned, for he is industrious at all times and unusually adroit in catching insects, especially in the moth stage.

The Mocking-bird

(Mimus polyglottus) Mocking-bird family

Length—9 to 10 inches. About the size of the robin.

Male and Female—Gray above; wings and wedge-shaped tail brownish; upper wing feathers tipped with white; outer tail quills white, conspicuous in flight; chin white; underneath light gray, shading to whitish.

Range—Peculiar to torrid and temperate zones of two Americas.

Migrations—No fixed migrations; usually resident where seen.

Dusky, Gray, and Slate-colored

North of Delaware this commonest of Southern birds is all too rarely seen outside of cages, yet even in midwinter it is not unknown in Central Park, New York. This is the angel that it is said the catbird was before he fell from grace. Slim, neat, graceful, imitative, amusing, with a rich, tender song that only the thrush can hope to rival, and with an instinctive preference for the society of man, it is little wonder he is a favorite, caged or free. He is a most devoted parent, too, when the four or six speckled green eggs have produced as many mouths to be supplied with insects and berries.

In the Connecticut Valley, where many mocking-birds' nests have been found, year after year, they are all seen near the ground, and without exception are loosely, poorly constructed affairs of leaves, feathers, grass, and even rags.

With all his virtues, it must be added, however, that this charming bird is a sad tease. There is no sound, whether made by bird or beast about him, that he cannot imitate so clearly as to deceive every one but himself. Very rarely can you find a mocking-bird without intelligence and mischief enough to appreciate his ventriloquism. In Sidney Lanier's college note-book was found written this reflection: "A poet is the mocking-bird of the spiritual universe. In him are collected all the individual songs of all individual natures." Later in life, with the same thought in mind, he referred to the bird as "yon slim Shakespeare on the tree." His exquisite stanzas, "To Our Mocking-bird," exalt the singer with the immortals :

> " Trillets of humor,—shrewdest whistle-wit—
> Contralto cadences of grave desire,
> Such as from off the passionate Indian pyre
> Drift down through sandal-odored flames that split
> About the slim young widow, who doth sit
> And sing above,—midnights of tone entire,—
> Tissues of moonlight, shot with songs of fire ;—
> Bright drops of tune, from oceans infinite
> Of melody, sipped off the thin-edged wave
> And trickling down the beak,—discourses brave
> Of serious matter that no man may guess,—
> Good-fellow greetings, cries of light distress—
> All these but now within the house we heard :
> O Death, wast thou too deaf to hear the bird ?

JUNCO.
Life-size.

" Nay, Bird ; my gnet gainsays the Lord's best right.
The Lord was fain, at some late festal time,
That Keats should set all heaven's woods in rhyme,
And Thou in bird-notes. Lo, this tearful night
Methinks I see thee, fresh from Death's despite,
Perched in a palm-grove, wild with pantomime
O'er blissful companies couched in shady thyme.
Methinks I hear thy silver whistlings bright
Meet with the mighty discourse of the wise,—
'Till broad Beethoven, deaf no more, and Keats,
'Midst of much talk, uplift their smiling eyes
And mark the music of thy wood-conceits,
And half-way pause on some large courteous word,
And call thee ' Brother,' O thou heavenly Bird ! "

Junco

(Junco hyemalis) Finch family

Called also : SNOWBIRD ; SLATE-COLORED SNOWBIRD

Length—5.5 to 6.5 inches. About the size of the English sparrow

Male—Upper parts slate-colored; darkest on head and neck, which
    are sometimes almost black and marked like a cowl. Gray
    on breast, like a vest. Underneath white. Several outer tail
    feathers white, conspicuous in flight.

Female—Lighter gray, inclining to brown.

Range—North America. Not common in warm latitudes. Breeds
    in the Catskills and northern New England.

Migrations—September. April. Winter resident.

"Leaden skies above; snow below," is Mr. Parkhurst's suggestive description of this rather timid little neighbor, that is only
starved into familiarity. When the snow has buried seed and
berries, a flock of juncos, mingling sociably with the sparrows
and chickadees about the kitchen door, will pick up scraps of
food with an intimacy quite touching in a bird naturally rather
shy. Here we can readily distinguish these "little gray-robed
monks and nuns," as Miss Florence Merriam calls them.

They are trim, sprightly, sleek, and even natty; their disposi-
tions are genial and vivacious, not quarrelsome, like their sparrow
cousins, and what is perhaps best about them, they are birds we
may surely depend upon seeing in the winter months. A few
come forth in September, migrating at night from the deep

woods of the north, where they have nested and moulted during the summer ; but not until frost has sharpened the air are large numbers of them seen. Rejoicing in winter, they nevertheless do not revel in the deep and fierce arctic blasts, as the snowflakes do, but take good care to avoid the open pastures before the hard storms overtake them.

Early in the spring their song is sometimes heard before they leave us to woo and to nest in the north. Mr. Bicknell describes it as "a crisp call-note, a simple trill, and a faint, whispered warble, usually much broken, but not without sweetness."

White-breasted Nuthatch

(Sitta carolinensis) Nuthatch family

Called also : TREE-MOUSE ; DEVIL DOWNHEAD

Length—5.5 to 6 inches. A trifle smaller than the English sparrow.
Male and Female—Upper parts slate-color. Top of head and nape black. Wings dark slate, edged with black, that fades to brown. Tail feathers brownish black, with white bars. Sides of head and underneath white, shading to pale reddish under the tail. (Female's head leaden.) Body flat and compact. Bill longer than head.
Range—British provinces to Mexico. Eastern United States.
Migrations—October. April. Common resident. Most prominent in winter.

> " Shrewd little haunter of woods all gray,
> Whom I meet on my walk of a winter day—
> You're busy inspecting each cranny and hole
> In the ragged bark of yon hickory bole ;
> You intent on your task, and I on the law
> Of your wonderful head and gymnastic claw !
>
> The woodpecker well may despair of this feat—
> Only the fly with you can compete !
> So much is clear ; but I fain would know
> How you can so reckless and fearless go,
> Head upward, head downward, all one to you,
> Zenith and nadir the same in your view ? "
>
> —*Edith M. Thomas.*

Could a dozen lines well contain a fuller description or more apt characterization of a bird than these "To a Nuthatch"?

WHITE-BREASTED NUT HATCH.
Life-size.

With more artless inquisitiveness than fear, this lively little acrobat stops his hammering or hatcheting at your approach, and stretching himself out from the tree until it would seem he must fall off, he peers down at you, head downward, straight into your upturned opera-glasses. If there is too much snow on the upper side of a branch, watch how he runs along underneath it like a fly, busily tapping the bark, or adroitly breaking the decayed bits with his bill, as he searches for the spider's eggs, larvæ, etc., hidden there; yet somehow, between mouthfuls, managing to call out his cheery *quank! quank! hank! hank!*

Titmice and nuthatches, which have many similar characteristics, are often seen in the most friendly hunting parties on the same tree. A pine woods is their dearest delight. There, as the mercury goes down, their spirits only seem to go up higher. In the spring they have been thought by many to migrate in flocks, whereas they are only retreating with their relations away from the haunts of men to the deep, cool woods, where they nest. With infinite patience the nuthatch executes a hole in a tree, lining it with feathers and moss, and often depositing as many as ten white eggs (speckled with red and lilac) for a single brood.

Red-breasted Nuthatch

(Sitta canadensis) Nuthatch family

Called also: CANADA NUTHATCH

Length—4 to 4.75 inches. One-third smaller than the English sparrow.

Male—Lead-colored above; brownish on wings and tail. Head, neck, and stripe passing through eye to shoulder, black. Frontlet, chin, and shoulders white; also a white stripe over eye, meeting on brow. Under parts light, rusty red. Tail feathers barred with white near end, and tipped with pale brown.

Female—Has crown of brownish black, and is lighter beneath than male.

Range—Northern parts of North America. Not often seen south of the most northerly States.

Migrations—November. April. Winter resident.

The brighter coloring of this tiny, hardy bird distinguishes it from the other and larger nuthatch, with whom it is usually seen, for the winter birds have a delightfully social manner, so that a colony of these Free masons is apt to contain not only both kinds of nuthatches and chickadees, but kinglets and brown creepers as well. It shares the family habit of walking about the trees, head downward, and running along the under side of limbs like a fly. By Thanksgiving Day the *quank! quank!* of the white-breasted species is answered by the *tai-tai-tait!* of the red-breasted cousin in the orchard, where the family party is celebrating with an elaborate *menu* of slugs, insects' eggs, and oily seeds from the evergreen trees.

For many years this nuthatch, a more northern species than the white-breasted bird, was thought to be only a spring and autumn visitor, but latterly it is credited with habits like its congener's in nearly every particular.

Loggerhead Shrike

(Lanius ludovicianus) Shrike family

Length—8.5 to 9 inches. A little smaller than the robin.

Male and Female—Upper parts gray; narrow black line across forehead, connecting small black patches on sides of head at base of bill. Wings and tail black, plentifully marked with white, the outer tail feathers often being entirely white and conspicuous in flight. Underneath white or very light gray. Bill hooked and hawk-like.

Range—Eastern United States to the plains.

Migrations—May. October. Summer resident.

It is not easy, even at a slight distance, to distinguish the loggerhead from the Northern shrike. Both have the pernicious habit of killing insects and smaller birds and impaling them on thorns; both have the peculiarity of flying, with strong, vigorous flight and much wing-flapping, close along the ground, then suddenly rising to a tree, on the lookout for prey. Their harsh, unmusical call-notes are similar too, and their hawk-like method of dropping suddenly upon a victim on the ground below is identical. Indeed, the same description very nearly answers for both birds. But there is one very important difference. While the Northern shrike is a winter visitor, the loggerhead, being his Southern counterpart, does not arrive until after the frost is out of the ground, and he can be sure of a truly warm welcome. A lesser

NORTHERN SHRIKE.
⅔ Life-size.

distinction between the only two representatives of the shrike family that frequent our neighborhood—and they are two too many—is in the smaller size of the loggerhead and its lighter-gray plumage. But as both these birds select some high, commanding position, like a distended branch near the tree-top, a cupola, house-peak, lightning-rod, telegraph wire, or weather-vane, the better to detect a passing dinner, it would be quite impossible at such a distance to know which shrike was sitting up there silently plotting villainies, without remembering the season when each may be expected.

Northern Shrike

(Lanius borealis) Shrike family

Called also : BUTCHER-BIRD ; NINE-KILLER

Length—9.5 to 10.5 inches. About the size of the robin.

Male—Upper parts slate-gray; wing quills and tail black, edged and tipped with white, conspicuous in flight; a white spot on centre of outer wing feathers. A black band runs from bill, through eye to side of throat. Light gray below, tinged with brownish, and faintly marked with waving lines of darker gray. Bill hooked and hawk-like.

Female—With eye-band more obscure than male's, and with more distinct brownish cast on her plumage.

Range—Northern North America. South in winter to middle portion of United States.

Migrations—November. April. A roving winter resident.

" Matching the bravest of the brave among birds of prey in deeds of daring, and no less relentless than reckless, the shrike compels that sort of deference, not unmixed with indignation, we are accustomed to accord to creatures of seeming insignificance whose exploits demand much strength, great spirit, and insatiate love for carnage. We cannot be indifferent to the marauder who takes his own wherever he finds it—a feudal baron who holds his own with undisputed sway—and an ogre whose victims are so many more than he can eat, that he actually keeps a private graveyard for the balance." Who is honestly able to give the shrikes a better character than Dr. Coues, just quoted? A few offer them questionable defence by recording the large numbers of English sparrows they kill in a season, as if wanton carnage were ever justifiable.

Not even a hawk itself can produce the consternation among a flock of sparrows that the harsh, rasping voice of the butcher-bird creates, for escape they well know to be difficult before the small ogre swoops down upon his victim, and carries it off to impale it on a thorn or frozen twig, there to devour it later piecemeal. Every shrike thus either impales or else hangs up, as a butcher does his meat, more little birds of many kinds, field-mice, grasshoppers, and other large insects than it can hope to devour in a week of bloody orgies. Field-mice are perhaps its favorite diet, but even snakes are not disdained.

· More contemptible than the actual slaughter of its victims, if possible, is the method by which the shrike often lures and sneaks upon his prey. Hiding in a clump of bushes in the meadow or garden, he imitates with fiendish cleverness the call-notes of little birds that come in cheerful response, hopping and flitting within easy range of him. His bloody work is finished in a trice. Usually, however, it must be owned, the shrike's hunting habits are the reverse of sneaking. Perched on a point of vantage on some tree-top or weather-vane, his hawk-like eye can detect a grasshopper going through the grass fifty yards away.

What is our surprise when some fine warm day in March, just before our butcher, ogre, sneak, and fiend leaves us for colder regions, to hear him break out into song ! Love has warmed even his cold heart, and with sweet, warbled notes on the tip of a beak that but yesterday was reeking with his victim's blood, he starts for Canada, leaving behind him the only good impression he has made during a long winter's visit.

Bohemian Waxwing

(*Ampelis garrulus*) Waxwing family

Called also : BLACK-THROATED WAXWING ; LAPLAND WAXWING ; SILKTAIL

Length—8 to 9.5 inches. A little smaller than the robin.
Male and Female—General color drab, with faint brownish wash above, shading into lighter gray below. Crest conspicuous,

88

being nearly an inch and a half in length; rufous at the base, shading into light gray above. Velvety-black forehead, chin, and line through the eye. Wings grayish brown, with very dark quills, which have two white bars; the bar at the edge of the upper wing coverts being tipped with red sealing-wax-like points, that give the bird its name. A few wing feathers tipped with yellow on outer edge. Tail quills dark brown, with yellow band across the end, and faint red streaks on upper and inner sides.

Range—Northern United States and British America. Most common in Canada and northern Mississippi region.

Migrations—Very irregular winter visitor.

When Charles Bonaparte, Prince of Canino, who was the first to count this common waxwing of Europe and Asia among the birds of North America, published an account of it in his "Synopsis," it was considered a very rare bird indeed. It may be these waxwings have greatly increased, but however uncommon they may still be considered, certainly no one who had ever seen a flock containing more than a thousand of them, resting on the trees of a lawn within sight of New York City, as the writer has done, could be expected to consider the birds "very rare."

The Bohemian waxwing, like the only other member of the family that ever visits us, the cedar-bird, is a roving gipsy. In Germany they say seven years must elapse between its visitations, which the superstitious old cronies are wont to associate with woful stories of pestilence—just such tales as are resurrected from the depths of morbid memories here when a comet reappears or the seven-year locust ascends from the ground.

The goings and comings of these birds are certainly most erratic and infrequent; nevertheless, when hunger drives them from the far north to feast upon the juniper and other winter berries of our Northern States, they come in enormous flocks, making up in quantity what they lack in regularity of visits and evenness of distribution.

Surely no bird has less right to be associated with evil than this mild waxwing. It seems the very incarnation of peace and harmony. Part of a flock that has lodged in a tree will sit almost motionless for hours and whisper in softly hissed twitterings, very much as a company of Quaker ladies, similarly dressed, might sit at yearly meeting. Exquisitely clothed in silky-gray feathers that no berry juice is ever permitted to stain, they are

89

dainty, gentle, aristocratic-looking birds, a trifle heavy and indolent, perhaps, when walking on the ground or perching; but as they fly in compact squads just above the tree-tops their flight is exceedingly swift and graceful.

Bay-breasted Warbler

(Dendroica castanea) Wood Warbler family

Length—5.25 to 5.75 inches. A little smaller than the English sparrow.

Male—Crown, chin, throat, upper breast, and sides dull chestnut. Forehead, sides of head, and cheeks black. Above olive-gray, streaked with black. Underneath buffy. Two white wing-bars. Outer tail quills with white patches on tips. Cream-white patch on either side of neck.

Female—Has more greenish-olive above.

Range—Eastern North America, from Hudson's Bay to Central America. Nests north of the United States. Winters in tropical limit of range.

Migrations—May. September. Rare migrant.

The chestnut breast of this capricious little visitor makes him look like a diminutive robin. In spring, when these warblers are said to take a more easterly route than the one they choose in autumn to return by to Central America, they may be so suddenly abundant that the fresh green trees and shrubbery of the garden will contain a dozen of the busy little hunters. Another season they may pass northward either by another route or leave your garden unvisited; and perhaps the people in the very next town may be counting your rare bird common, while it is simply perverse.

Whether common or rare, before your acquaintance has had time to ripen into friendship, away go the freaky little creatures to nest in the tree-tops of the Canadian coniferous forests.

Chestnut-sided Warbler

(Dendroica pennsylvanica) Wood Warbler family

Called also: BLOODY-SIDED WARBLER

Length—About 5 inches. Over an inch shorter than the English sparrow.

Male—Top of head and streaks in wings yellow. A black line

BOHEMIAN WAXWING.

running through the eye and round back of crown, and a black spot in front of eye, extending to cheeks. Ear coverts, chin, and underneath white. Back greenish gray and slate, streaked with black. Sides of bird chestnut. Wings, which are streaked with black and yellow, have yellowish-white bars. Very dark tail with white patches on inner vanes of the outer quills.

Female—Similar, but duller. Chestnut sides are often scarcely apparent.

Range—Eastern North America, from Manitoba and Labrador to the tropics, where it winters.

Migrations—May. September. Summer resident, most common in migrations.

In the Alleghanies, and from New Jersey and Illinois northward, this restless little warbler nests in the bushy borders of woodlands and the undergrowth of the woods, for which he forsakes our gardens and orchards after a very short visit in May. While hopping over the ground catching ants, of which he seems to be inordinately fond, or flitting actively about the shrubbery after grubs and insects, we may note his coat of many colors—patchwork in which nearly all the warbler colors are curiously combined. With drooped wings that often conceal the bird's chestnut sides, which are his chief distinguishing mark, and with tail erected like a redstart's, he hunts incessantly. Here in the garden he is as refreshingly indifferent to your interest in him as later in his breeding haunts he is shy and distrustful. His song is bright and animated, like that of the yellow warbler.

Golden-winged Warbler

(Helminthophila chrysoptera) Wood Warbler family

Length—About 5 inches. More than an inch shorter than the English sparrow.

Male—Yellow crown and yellow patches on the wings. Upper parts bluish gray, sometimes tinged with greenish. Stripe through the eye and throat black. Sides of head, chin, and line over the eye white. Underneath white, grayish on sides. A few white markings on outer tail feathers.

Female—Crown duller ; gray where male is black, with olive upper parts and grayer underneath.

Range—From Canadian border to Central America, where it winters.
Migrations—May. September. Summer resident.

After one has seen a golden-winged warbler fluttering hither and thither about the shrubbery of a park within sight and sound of a great city's distractions and with blissful unconcern of them all, partaking of a hearty lunch of insects that infest the leaves before one's eyes, one counts the bird less rare and shy than one has been taught to consider it. Whoever looks for a warbler with gaudy yellow wings will not find the golden-winged variety. His wings have golden patches only, and while these are distinguishing marks, they are scarcely prominent enough features to have given the bird the rather misleading name he bears. But, then, most warblers' names are misleading. They serve their best purpose in cultivating patience and other gentle virtues in the novice.

Such habits and choice of haunts as characterize the blue-winged warbler are also the golden-winged's. But their voices are quite different, the former's being sharp and metallic, while the latter's *ʒee, ʒee, ʒee* comes more lazily and without accent.

Myrtle Warbler

(Dendroica coronata) Wood Warbler family

Called also: YELLOW-RUMPED WARBLER ; MYRTLE-BIRD; YELLOW-CROWNED WARBLER

Length—5 to 5.5 inches. About an inch smaller than the English sparrow.
Male—In summer plumage: A yellow patch on top of head, lower back, and either side of the breast. Upper parts bluish slate, streaked with black. Upper breast black ; throat white; all other under parts whitish, streaked with black. Two white wing-bars, and tail quills have white spots near the tip. *In winter:* Upper parts olive-brown, streaked with black; the yellow spot on lower back the only yellow mark remaining. Wing-bars grayish.
Female—Resembles male in winter plumage.
Range—Eastern North America. Occasional on Pacific slope. Summers from Minnesota and northern New England northward to Fur Countries. Winters from Middle States south-

ward into Central America; a few often remaining at the
northern United States all the winter.
Migrations—April. October. November. Also, but more rarely,
a winter resident.

The first of the warblers to arrive in the spring and the last
to leave us in the autumn, some even remaining throughout the
northern winter, the myrtle warbler, next to the summer yellowbird, is the most familiar of its multitudinous kin. Though we
become acquainted with it chiefly in the migrations, it impresses
us by its numbers rather than by any gorgeousness of attire. The
four yellow spots on crown, lower back, and sides are its distinguishing marks; and in the autumn these marks have dwindled
to only one, that on the lower back or rump. The great difficulty experienced in identifying any warbler is in its restless habit
of flitting about.

For a few days in early May we are forcibly reminded of the
Florida peninsula, which fairly teems with these birds ; they
become almost superabundant, a distraction during the precious
days when the rarer species are quietly slipping by, not to return
again for a year, perhaps longer, for some warblers are notoriously
irregular in their routes north and south, and never return by the
way they travelled in the spring.

But if we look sharply into every group of myrtle warblers,
we are quite likely to discover some of their dainty, fragile cousins that gladly seek the escort of birds so fearless as they. By
the last of May all the warblers are gone from the neighborhood
except the constant little summer yellowbird and redstart.

In autumn, when the myrtle warblers return after a busy
enough summer passed in Canadian nurseries, they chiefly haunt
those regions where juniper and bay-berries abound. These latter
(Myrica cerifera), or the myrtle wax-berries, as they are sometimes called, and which are the bird's favorite food, have given it
their name. Wherever the supply of these berries is sufficient to
last through the winter, there it may be found foraging in the
scrubby bushes. Sometimes driven by cold and hunger from
the fields, this hardiest member of a family that properly belongs
to the tropics, seeks shelter and food close to the outbuildings
on the farm.

Parula Warbler

(Compsothlypis americana) Wood Warbler family

Called also: BLUE YELLOW-BACKED WARBLER

Length—4.5 to 4.75 inches. About an inch and a half shorter than the English sparrow.

Male and Female—Slate-colored above, with a greenish-yellow or bronze patch in the middle of the back. Chin, throat, and breast yellow. A black, bluish, or rufous band across the breast, usually lacking in female. Underneath white, sometimes marked with rufous on sides, but these markings are variable. Wings have two white patches; outer tail feathers have white patch near the end.

Range—Eastern North America. Winters from Florida southward.

Migrations—April. October. Summer resident.

Through an open window of an apartment in the very heart of New York City, a parula warbler flew this spring of 1897, surely the daintiest, most exquisitely beautiful bird visitor that ever voluntarily lodged between two brick walls.

A number of such airy, tiny beauties flitting about among the blossoms of the shrubbery on a bright May morning and swaying on the slenderest branches with their inimitable grace, is a sight that the memory should retain into old age. They seem the very embodiment of life, joy, beauty, grace; of everything lovely that birds by any possibility could be. Apparently they are wafted about the garden; they fly with no more effort than a dainty lifting of the wings, as if to catch the breeze, that seems to lift them as it might a bunch of thistledown. They go through a great variety of charming posturings as they hunt for their food upon the blossoms and tender fresh twigs, now creeping like a nuthatch along the bark and peering into the crevices, now gracefully swaying and balancing like a goldfinch upon a slender, pendent stem. One little sprite pauses in its hunt for the insects to raise its pretty head and trill a short and wiry song.

But the parula warbler does not remain long about the gardens and orchards, though it will not forsake us altogether for the Canadian forests, where most of its relatives pass the summer. It retreats only to the woods near the water, if may be, or to just as close a counterpart of a swampy southern woods, where the

Spanish or Usnea " moss " drapes itself over the cypresses, as it can find here at the north. Its rarely beautiful nest, that hangs suspended from a slender branch very much like the Baltimore oriole's, is so woven and festooned with this moss that its concealment is perfect.

Black-throated Blue Warbler

(Dendroica cærulescens) Wood Warbler family

Length—5.30 inches. About an inch shorter than the English sparrow.

Male—Slate-color, not blue above ; lightest on forehead and darkest on lower back. Wings and tail edged with bluish. Cheeks, chin, throat, upper breast, and sides black. Breast and underneath white. White spots on wings, and a little white on tail.

Female—Olive-green above ; underneath soiled yellow. Wing-spots inconspicuous. Tail generally has a faint bluish tinge.

Range—Eastern North America, from Labrador to tropics, where it winters.

Migrations—May. September. Usually a migrant only in the United States.

Whoever looks for this beautifully marked warbler among the bluebirds, will wish that the man who named him had possessed a truer eye for color. But if the name so illy fits the bright slate-colored male, how grieved must be his little olive-and-yellow mate to answer to the name of black-throated blue warbler when she has neither a black throat nor a blue feather! It is not easy to distinguish her as she flits about the twigs and leaves of the garden in May or early autumn, except as she is seen in company with her husband, whose name she has taken with him for better or for worse. The white spot on the wings should always be looked for to positively identify this bird.

Before flying up to a twig to peck off the insects, the birds have a pretty vireo trick of cocking their heads on one side to investigate the quantity hidden underneath the leaves. They seem less nervous and more deliberate than many of their restless family.

Most warblers go over the Canada border to nest, but there are many records of the nests of this species in the Alleghanies as far south as Georgia, in the Catskills, in Connecticut, northern

95

Minnesota and Michigan. Laurel thickets and moist undergrowth of woods in the United States, and more commonly pine woods in Canada, are the favorite nesting haunts. A sharp *zip, zip,* like some midsummer insect's noise, is the bird's call-note, but its love-song, *zee, zee, zee,* or *twee, twea, twea-e-e,* as one authority writes it, is only rarely heard in the migrations. It is a languid, drawling little strain, with an upward slide that is easily drowned in the full bird chorus of May.

BLUE AND BLUISH BIRDS

Bluebird
Indigo Bunting
Belted Kingfisher
Blue Jay
Blue Grosbeak
Barn Swallow
Cliff Swallow
Mourning Dove
Blue-gray Gnatcatcher

Look also among Slate-colored Birds in preceding group, particularly among the Warblers there, or in the group of Birds conspicuously Yellow and Orange.

BLUE BIRD.
Life-size.

BLUE AND BLUISH BIRDS

The Bluebird

(Sialia sialis) Thrush family

Called also : BLUE ROBIN

Length—7 inches. About an inch longer than the English sparrow.

Male—Upper parts, wings, and tail bright blue, with rusty wash in autumn. Throat, breast, and sides cinnamon-red. Underneath white.

Female—Has duller blue feathers, washed with gray, and a paler breast than male.

Range—North America, from Nova Scotia and Manitoba to Gulf of Mexico. Southward in winter from Middle States to Bermuda and West Indies.

Migrations—March. November. Summer resident. A few sometimes remain throughout the winter.

With the first soft, plaintive warble of the bluebirds early in March, the sugar camps, waiting for their signal, take on a bustling activity; the farmer looks to his plough; orders are hurried off to the seedsmen; a fever to be out of doors seizes one: spring is here. Snowstorms may yet whiten fields and gardens, high winds may howl about the trees and chimneys, but the little blue heralds persistently proclaim from the orchard and garden that the spring procession has begun to move. *Tru-al-ly, tru-al-ly,* they sweetly assert to our incredulous ears.

The bluebird is not always a migrant, except in the more northern portions of the country. Some representatives there are always with us, but the great majority winter south and drop out of the spring procession on its way northward, the males a little ahead of their mates, which show housewifely instincts immediately after their arrival. A pair of these rather undemonstrative, matter-of-fact lovers go about looking for some deserted woodpecker's hole in the orchard, peering into cavities in the fence-rails, or into the bird-houses that, once set up in the old-fashioned gardens for their special benefit, are now appropriated too often by the ubiquitous sparrow. Wrens they can readily dispossess of an attractive tenement, and do. With a temper as heavenly as the color of their feathers, the bluebird's sense of justice is not always so adorable. But sparrows unnerve them into cowardice. The comparatively infrequent nesting of the bluebirds about our homes at the present time is one of the most deplorable results of unrestricted sparrow immigration. Formerly they were the commonest of bird neighbors.

Nest-building is not a favorite occupation with the bluebirds, that are conspicuously domestic none the less. Two, and even three, broods in a season fully occupy their time. As in most cases, the mother-bird does more than her share of the work. The male looks with wondering admiration at the housewifely activity, applauds her with song, feeds her as she sits brooding over the nestful of pale greenish-blue eggs, but his adoration of her virtues does not lead him into emulation.

> " Shifting his light load of song,
> From post to post along the cheerless fence,"

Lowell observed that he carried his duties quite as lightly.

When the young birds first emerge from the shell they are almost black; they come into their splendid heritage of color by degrees, lest their young heads might be turned. It is only as they spread their tiny wings for their first flight from the nest that we can see a few blue feathers.

With the first cool days of autumn the bluebirds collect in flocks, often associating with orioles and kingbirds in sheltered, sunny places where insects are still plentiful. Their steady, undulating flight now becomes erratic as they take food on the wing— a habit that they may have learned by association with the kingbirds, for they have also adopted the habit of perching upon some conspicuous lookout and then suddenly launching out into the air for a passing fly and returning to their perch. Long after their associates have gone southward, they linger like the last leaves on the tree. It is indeed "good-bye to summer" when the bluebirds withdraw their touch of brightness from the dreary November landscape.

The bluebirds from Canada and the northern portions of New

INDIGO BIRD.
Life-size.

England and New York migrate into Virginia and the Carolinas;
the birds from the Middle States move down into the Gulf States
to pass the winter. It was there that countless numbers were
cut off by the severe winter of 1894-95, which was so severe in
that section.

Indigo Bunting

(Passerina cyanea) Finch family

Called also : INDIGO BIRD

Length—5.5 to 6 inches. Smaller than the English sparrow, or
    the size of a canary.

Male—In certain lights rich blue, deepest on head. In another
    light the blue feathers show verdigris tints. Wings, tail, and
    lower back with brownish wash, most prominent in autumn
    plumage. Quills of wings and tail deep blue, margined with
    light.

Female—Plain sienna-brown above. Yellowish on breast and
    shading to white underneath, and indistinctly streaked.
    Wings and tail darkest, sometimes with slight tinge of blue
    in outer webs and on shoulders.

Range—North America, from Hudson Bay to Panama. Most
    common in eastern part of United States. Winters in
    Central America and Mexico.

Migrations—May. September. Summer resident.

The "glowing indigo" of this tropical-looking visitor that
so delighted Thoreau in the Walden woods, often seems only the
more intense by comparison with the blue sky, against which it
stands out in relief as the bird perches singing in a tree-top.
What has this gaily dressed, dapper little cavalier in common
with his dingy sparrow cousins that haunt the ground and delight in dust-baths, leaving their feathers no whit more dingy
than they were before, and in temper, as in plumage, suggesting
more of earth than of heaven? Apparently he has nothing, and
yet the small brown bird in the roadside thicket, which you have
misnamed a sparrow, not noticing the glint of blue in her shoulders and tail, is his mate. Besides the structural resemblances,
which are, of course, the only ones considered by ornithologists
in classifying birds, the indigo buntings have several sparrow-
like traits. They feed upon the ground, mainly upon seeds of
grasses and herbs, with a few insects interspersed to give relish

to the grain ; they build grassy nests in low bushes or tall, rank grass ; and their flight is short and labored. Borders of woods, roadside thickets, and even garden shrubbery, with open pasture lots for foraging grounds near by, are favorite haunts of these birds, that return again and again to some preferred spot. But however close to our homes they build theirs, our presence never ceases to be regarded by them with anything but suspicion, not to say alarm. Their metallic *cheep, cheep,* warns you to keep away from the little blue-white eggs, hidden away securely in the bushes ; and the nervous tail twitchings and jerkings are pathetic to see. Happily for the safety of their nest, the brooding mother has no tell-tale feathers to attract the eye. Dense foliage no more conceals the male bird's brilliant coat than it can the tanager's or oriole's.

With no attempt at concealment, which he doubtless understands would be quite impossible, he chooses some high, conspicuous perch to which he mounts by easy stages, singing as he goes ; and there begins a loud and rapid strain that promises much, but growing weaker and weaker, ends as if the bird were either out of breath or too weak to finish. Then suddenly he begins the same song over again, and keeps up this continuous performance for nearly half an hour. The noonday heat of an August day that silences nearly every other voice, seems to give to the indigo bird's only fresh animation and timbre.

The Belted Kingfisher

(Ceryle alcyon) Kingfisher family

Called also : THE HALCYON

Length—12 to 13 inches. About one-fourth as large again as the robin.

Male—Upper part grayish blue, with prominent crest on head reaching to the nape. A white spot in front of the eye. Bill longer than the head, which is large and heavy. Wings and the short tail minutely speckled and marked with broken bands of white. Chin, band around throat, and underneath white. Two bluish bands across the breast and a bluish wash on sides.

Female—Female and immature specimens have rufous bands where the adult male's are blue. Plumage of both birds oily.

KINGFISHER.
½ Life-size.

Range—North America, except where the Texan kingfisher
replaces it in a limited area in the Southwest. Common from
Labrador to Florida, east and west. Winters chiefly from
Virginia southward to South America.
Migrations—March. December. Common summer resident.
Usually a winter resident also.

If the kingfisher is not so neighborly as we could wish, or as
he used to be, it is not because he has grown less friendly, but
because the streams near our homes are fished out. Fish he
must and will have, and to get them nowadays it is too often
necessary to follow the stream back through secluded woods to
the quiet waters of its source: a clear, cool pond or lake whose
scaly inmates have not yet learned wisdom at the point of the
sportsman's fly.

In such quiet haunts the kingfisher is easily the most conspicuous object in sight, where he perches on some dead or projecting branch over the water, intently watching for a dinner that
is all unsuspectingly swimming below. Suddenly the bird drops
—dives ; there is a splash, a struggle, and then the "lone fisherman" returns triumphant to his perch, holding a shining fish in his
beak. If the fish is small it is swallowed at once, but if it is large
and bony it must first be killed against the branch. A few sharp
knocks, and the struggles of the fish are over, but the kingfisher's
have only begun. How he gags and writhes, swallows his
dinner, and then, regretting his haste, brings it up again to try
another wider avenue down his throat ! The many abortive
efforts he makes to land his dinner safely below in his stomach,
his grim contortions as the fishbones scratch his throat-lining on
their way down and up again, force a smile in spite of the bird's
evident distress. It is small wonder he supplements his fish diet
with various kinds of the larger insects, shrimps, and fresh-water
mollusks.

Flying well over the tree-tops or along the waterways, the
kingfisher makes the woodland echo with his noisy rattle, that
breaks the stillness like a watchman's at midnight. It is, perhaps, the most familiar sound heard along the banks of the inland
rivers. No love or cradle song does he know. Instead of soften-
ing and growing sweet, as the voices of most birds do in the
nesting season, the endearments uttered by a pair of mated kingfishers are the most strident, rattly shrieks ever heard by lovers.

It sounds as if they were perpetually quarrelling, and yet they are really particularly devoted.

The nest of these birds, like the bank swallow's, is excavated in the face of a high bank, preferably one that rises from a stream; and at about six feet from the entrance of the tunnel six or eight clear, shining white eggs are placed on a curious nest. All the fish-bones and scales that, being indigestible, are disgorged in pellets by the parents, are carefully carried to the end of the tunnel to form a prickly cradle for the unhappy fledglings. Very rarely a nest is made in the hollow trunk of a tree; but wherever the home is, the kingfishers become strongly attached to it, returning again and again to the spot that has cost them so much labor to excavate. Some observers have accused them of appropriating the holes of the water-rats.

In ancient times of myths and fables, kingfishers or halcyons were said to build a floating nest on the sea, and to possess some mysterious power that calmed the troubled waves while the eggs were hatching and the young birds were being reared, hence the term "halcyon days," meaning days of fair weather

Blue Jay

(Cyanocitta cristata) Crow and Jay family

Length—11 to 12 inches. A little larger than the robin.

Male and Female—Blue above. Black band around the neck, joining some black feathers on the back. Under parts dusky white. Wing coverts and tail bright blue, striped transversely with black. Tail much rounded. Many feathers edged and tipped with white. Head finely crested ; bill, tongue, and legs black.

Range—Eastern coast of North America to the plains, and from northern Canada to Florida and eastern Texas.

Migrations—Permanent resident. Although seen in flocks moving southward or northward, they are merely seeking happier hunting grounds, not migrating.

No bird of finer color or presence sojourns with us the year round than the blue jay. In a peculiar sense his is a case of "beauty covering a multitude of sins." Among close students of bird traits, we find none so poor as to do him reverence. Dishonest, cruel, inquisitive, murderous, voracious, villainous, are

BLUE JAY.
⅔ Life-size.

some of the epithets applied to this bird of exquisite plumage. Emerson, however, has said in his defence he does "more good than harm," alluding, no doubt, to his habit of burying nuts and hard seeds in the ground, so that many a waste place is clothed with trees and shrubs, thanks to his propensity and industry.

He is mischievous as a small boy, destructive as a monkey, deft at hiding as a squirrel. He is unsociable and unamiable, disliking the society of other birds. His harsh screams, shrieks, and most aggressive and unmusical calls seem often intended maliciously to drown the songs of the sweet-voiced singers.

From April to September, the breeding and moulting season, the blue jays are almost silent, only sallying forth from the woods to pillage and devour the young and eggs of their more peaceful neighbors. In a bulky nest, usually placed in a tree-crotch high above our heads, from four to six eggs, olive-gray with brown spots, are laid and most carefully tended.

Notwithstanding the unlovely characteristics of the blue jay, we could ill spare the flash of color, like a bit of blue sky dropped from above, which is so rare a tint even in our land, that we number not more than three or four true blue birds, and in England, it is said, there is none.

Blue Grosbeak

(*Guiraca cærulea*) Finch family

Length—7 inches. About an inch larger than the English sparrow.

Male—Deep blue, dark, and almost black on the back; wings and tail black, slightly edged with blue, and the former marked with bright chestnut. Cheeks and chin black. Bill heavy and bluish.

Female—Grayish brown above, sometimes with bluish tinge on head, lower back, and shoulders. Wings dark olive-brown, with faint buff markings; tail same shade as wings, but with bluish-gray markings. Underneath brownish cream-color, the breast feathers often blue at the base.

Range—United States, from southern New England westward to the Rocky Mountains and southward into Mexico and beyond. Most common in the Southwest. Rare along the Atlantic seaboard.

Migrations—May. September. Summer resident.

This beautiful but rather shy and solitary bird occasionally

wanders eastward to rival the bluebird and the indigo bunting
in their rare and lovely coloring, and eclipse them both in song.
Audubon, we remember, found the nest in New Jersey. Pennsylvania is still favored with one now and then, but it is in the
Southwest only that the blue grosbeak is as common as the
evening grosbeak is in the Northwest. Since rice is its favorite
food, it naturally abounds where that cereal grows. Seeds and
kernels of the hardest kinds, that its heavy, strong beak is well
adapted to crack, constitute its diet when it strays beyond the
rice-fields.

Possibly the heavy bills of all the grosbeaks make them look
stupid whether they are or not—a characteristic that the blue grosbeak's habit of sitting motionless with a vacant stare many minutes at a time unfortunately emphasizes.

When seen in the roadside thickets or tall weeds, such as the
field sparrow chooses to frequent, it shows little fear of man unless actually approached and threatened, but whether this fearlessness comes from actual confidence or stupidity is by no means
certain. Whatever the motive of its inactivity, it accomplishes an
end to be desired by the cleverest bird ; its presence is almost
never suspected by the passer-by, and its grassy nest on a tree-
branch, containing three or four pale bluish-white eggs, is never
betrayed by look or sign to the marauding small boy.

Barn Swallow

(Chelidon erythrogaster) Swallow family

Length—6.5 to 7 inches. A trifle larger than the English sparrow.
  Apparently considerably larger, because of its wide wing-
  spread.

Male—Glistening steel-blue shading to black above. Chin, breast,
  and underneath bright chestnut-brown and brilliant buff that
  glistens in the sunlight. A partial collar of steel-blue. Tail
  very deeply forked and slender.

Female—Smaller and paler, with shorter outer tail feathers, making the fork less prominent.

Range—Throughout North America. Winters in tropics of both
  Americas.

Migrations—April. September. Summer resident.

Any one who attempts to describe the coloring of a bird's
plumage knows how inadequate words are to convey a just idea

BARN SWALLOW.
⅔ Life-size.

of the delicacy, richness, and brilliancy of the living tints. But, happily, the beautiful barn swallow is too familiar to need description. Wheeling about our barns and houses, skimming over the fields, its bright sides flashing in the sunlight, playing "cross tag" with its friends at evening, when the insects, too, are on the wing, gyrating, darting, and gliding through the air, it is no more possible to adequately describe the exquisite grace of a swallow's flight than the glistening buff of its breast.

This is a typical bird of the air, as an oriole is of the trees and a sparrow of the ground. Though the swallow may often be seen perching on a telegraph wire, suddenly it darts off as if it had received a shock of electricity, and we see the bird in its true element.

While this swallow is peculiarly American, it is often confounded with its European cousin *Hirundo rustica* in noted ornithologies.

Up in the rafters of the barn, or in the arch of an old bridge that spans a stream, these swallows build their bracket-like nests of clay or mud pellets intermixed with straw. Here the noisy little broods pick their way out of the white eggs curiously spotted with brown and lilac that were all too familiar in the marauding days of our childhood.

Cliff Swallow

(Petrochelidon lunifrons) Swallow family

Called also : EAVE SWALLOW; CRESCENT SWALLOW; ROCKY MOUNTAIN SWALLOW

Length—6 inches. A trifle smaller than the English sparrow. Apparently considerably larger because of its wide wing-spread.

Male and Female—Steel-blue above, shading to blue-black on crown of head and on wings and tail. A brownish-gray ring around the neck. Beneath dusty white, with rufous tint. Crescent-like frontlet. Chin, throat, sides of head, and tail coverts rufous.

Range—North and South America. Winters in the tropics.

Migrations—Early April. Late September. Summer resident.

Not quite so brilliantly colored as the barn swallow, nor with tail so deeply forked, and consequently without so much grace in flying, and with a squeak rather than the really musical twitter of the gayer bird, the cliff swallow may be positively identified by the rufous feathers of its tail coverts, but more definitely by its crescent-shaped frontlet shining like a new moon; hence its specific Latin name from *luna* = moon, and *frons* = front.

Such great numbers of these swallows have been seen in the far West that the name of Rocky Mountain swallows is sometimes given to them; though however rare they may have been in 1824, when DeWitt Clinton thought he "discovered" them near Lake Champlain, they are now common enough in all parts of the United States.

In the West this swallow is wholly a cliff-dweller, but it has learned to modify its home in different localities. As usually seen, it is gourd-shaped, opened at the top, built entirely of mud pellets ("bricks without straw"), softly lined with feathers and wisps of grass, and attached by the larger part to a projecting cliff or eave.

Like all the swallows, this bird lives in colonies, and the clay-colored nests beneath the eaves of barns are often so close together that a group of them resembles nothing so much as a gigantic wasp's nest. It is said that when swallows pair they are mated for life; but, then, more is said about swallows than the most tireless bird-lover could substantiate. The tradition that swallows fly low when it is going to rain may be easily credited, because the air before a storm is usually too heavy with moisture for the winged insects, upon which the swallows feed, to fly high.

Mourning Dove

(Zenaidura macroura) Pigeon family

Called also: CAROLINA DOVE; TURTLE DOVE

Length—12 to 13 inches. About one-half as large again as the robin.

Male—Grayish brown or fawn-color above, varying to bluish gray. Crown and upper part of head greenish blue, with green and golden metallic reflections on sides of neck. A black spot under each ear. Forehead and breast reddish buff; lighter underneath. (General impression of color, bluish fawn.) Bill black, with tumid, fleshy covering; feet red; two middle tail feathers longest; all others banded with black

108

and tipped with ashy white.   Wing coverts sparsely spotted
with black.   Flanks and underneath the wings bluish.

Female—Duller and without iridescent reflections on neck.

Range—North America, from Quebec to Panama, and westward
to Arizona.   Most common in temperate climate, east of
Rocky Mountains.

Migrations—March.   November.   Common summer resident;
not migratory south of Virginia.

The beautiful, soft-colored plumage of this incessant and
rather melancholy love-maker is not on public exhibition.   To see
it we must trace the *a-coo-o, coo-o, coo-oo, coo-o* to its source in
the thick foliage in some tree in an out-of-the-way corner of the
farm, or to an evergreen near the edge of the woods.   The slow,
plaintive notes, more like a dirge than a love-song, penetrate to
a surprising distance.   They may not always be the same lovers
we hear from April to the end of summer, but surely the sound
seems to indicate that they are.   The dove is a shy bird, attached
to its gentle and refined mate with a devotion that has passed
into a proverb, but caring little or nothing for the society of other
feathered friends, and very little for its own kind, unless after the
nesting season has passed.   In this respect it differs widely from
its cousins, the wild pigeons, flocks of which, numbering many
millions, are recorded by Wilson and other early writers before
the days when netting these birds became so fatally profitable.

What the dove finds to adore so ardently in the "shiftless
housewife," as Mrs. Wright calls his lady-love, must pass the
comprehension of the phœbe, that constructs such an exquisite
home, or of a bustling, energetic Jenny wren, that "looketh well
to the ways of her household and eateth not the bread of idle-
ness."   She is a flabby, spineless bundle of flesh and pretty
feathers, gentle and refined in manners, but slack and incompe-
tent in all she does.   Her nest consists of a few loose sticks,
without rim or lining; and when her two babies emerge from
the white eggs, that somehow do not fall through or roll out of
the rickety lattice, their tender little naked bodies must suffer
from many bruises.   We are almost inclined to blame the incon-
siderate mother for allowing her offspring to enter the world
unclothed—obviously not her fault, though she is capable of just
such negligence.   Fortunate are the baby doves when their lazy
mother scatters her makeshift nest on top of one that a robin has

deserted, as she frequently does. It is almost excusable to take her young birds and rear them in captivity, where they invariably thrive, mate, and live happily, unless death comes to one, when the other often refuses food and grieves its life away.

In the wild state, when the nesting season approaches, both birds make curious acrobatic flights above the tree-tops; then, after a short sail in midair, they return to their perch. This appears to be their only giddiness and frivolity, unless a dust-bath in the country road might be considered a dissipation.

In the autumn a few pairs of doves show slight gregarious tendencies, feeding amiably together in the grain fields and retiring to the same roost at sundown.

Blue-gray Gnatcatcher

(Polioptila cærulea) Gnatcatcher family

Called also : SYLVAN FLYCATCHER

Length—4.5 inches. About two inches smaller than the English sparrow.

Male—Grayish blue above, dull grayish white below. Grayish tips on wings. Tail with white outer quills changing gradually through black and white to all black on centre quills. Narrow black band over the forehead and eyes. Resembles in manner and form a miniature catbird.

Female—More grayish and less blue, and without the black on head.

Range—United States to Canadian border on the north, the Rockies on the west, and the Atlantic States, from Maine to Florida ; most common in the Middle States. A rare bird north of New Jersey. Winters in Mexico and beyond.

Migrations—May. September. Summer resident.

In thick woodlands, where a stream that lazily creeps through the mossy, oozy ground attracts myriads of insects to its humid neighborhood, this tiny hunter loves to hide in the denser foliage of the upper branches. He has the habit of nervously flitting about from twig to twig of his relatives, the kinglets, but unhappily he lacks their social, friendly instincts, and therefore is rarely seen. Formerly classed among the warblers, then among the fly-catchers, while still as much a lover of flies, gnats, and mosquitoes as ever, his vocal powers have now won for him recognition

among the singing birds.   Some one has likened his voice to the
squeak of a mouse, and Nuttall says it is "scarcely louder," which
is all too true, for at a little distance it is quite inaudible.   But in
addition to the mouse-like call-note, the tiny bird has a rather
feeble but exquisitely finished song, so faint it seems almost as if
the bird were singing in its sleep.

If by accident you enter the neighborhood of its nest, you
soon find out that this timid, soft-voiced little creature can be
roused to rashness and make its presence disagreeable to ears and
eyes alike as it angrily darts about your unoffending head, pecking at your face and uttering its shrill squeak close to your very
ear-drums.   All this excitement is in defence of a dainty, lichen-
covered nest, whose presence you may not have even suspected
before, and of four or five bluish-white, speckled eggs well beyond reach in the tree-tops.

During the migrations the bird seems not unwilling to show
its delicate, trim little body, that has often been likened to a diminutive mocking-bird's, very near the homes of men.   Its graceful postures, its song and constant motion, are sure to attract
attention.   In Central Park, New York City, the bird is not
unknown.

BROWN, OLIVE OR GRAYISH BROWN, AND
BROWN AND GRAY SPARROWY BIRDS

House Wren
Carolina Wren
Winter Wren
Long-billed Marsh Wren
Short-billed Marsh Wren
Brown Thrasher
Wilson's Thrush or Veery
Wood Thrush
Hermit Thrush
Alice's Thrush
Olive-backed Thrush
Louisiana Water Thrush
Northern Water Thrush
Flicker
Meadowlark and Western
    Meadowlark
Horned Lark and Prairie
    Horned Lark
Pipit or Titlark
Whippoorwill
Nighthawk
Black-billed Cuckoo

Yellow-billed Cuckoo
Bank Swallow and Rough-
    winged Swallow
Cedar Bird
Brown Creeper
Pine Siskin
Smith's Painted Longspur
Lapland Longspur
Chipping Sparrow
English Sparrow
Field Sparrow
Fox Sparrow
Grasshopper Sparrow
Savanna Sparrow
Seaside Sparrow
Sharp-tailed Sparrow
Song Sparrow
Swamp Song Sparrow
Tree Sparrow
Vesper Sparrow
White-crowned Sparrow
White-throated Sparrow

See also winter plumage of the Bobolink, Goldfinch, and Myrtle Warbler. See females of Red-winged Blackbird, Rusty Blackbird, the Grackles, Bobolink, Cowbird, the Redpolls, Purple Finch, Chewink, Bluebird, Indigo Bunting, Baltimore Oriole, Cardinal, and of the Evening, the Blue, and the Rose-breasted Grosbeaks. See also Purple Finch, the Redpolls, Mourning Dove, Mocking-bird, Robin.

HOUSE WREN.
Life-size.

BROWN, OLIVE OR GRAYISH BROWN, AND BROWN AND GRAY SPARROWY BIRDS

House Wren

(Troglodytes aëdon) Wren family

Length—4.5 to 5 inches. Actually about one-fourth smaller than the English sparrow; apparently only half as large because of its erect tail.

Male and Female—Upper parts cinnamon-brown. Deepest shade on head and neck; lightest above tail, which is more rufous. Back has obscure, dusky bars ; wings and tail are finely barred. Underneath whitish, with grayish-brown wash and faint bands most prominent on sides.

Range—North America, from Manitoba to the Gulf. Most common in the United States, from the Mississippi eastward. Winters south of the Carolinas.

Migrations—April. October. Common summer resident.

Early some morning in April there will go off under your window that most delightful of all alarm-clocks—the tiny, friendly house wren, just returned from a long visit south. Like some little mountain spring that, having been imprisoned by winter ice, now bubbles up in the spring sunshine, and goes rippling along over the pebbles, tumbling over itself in merry cascades, so this little wren's song bubbles, ripples, cascades in a miniature torrent of ecstasy.

Year after year these birds return to the same nesting places : a box set up against the house, a crevice in the barn, a niche under the eaves; but once home, always home to them. The nest is kept scrupulously clean ; the house-cleaning, like the house-building and renovating, being accompanied by the cheeriest of songs, that makes the bird fairly tremble by its intensity. But however angelic the voice of the house wren, its temper can put to flight even the English sparrow. Need description go further ?

Six to eight minutely speckled, flesh-colored eggs suffice to keep the nervous, irritable parents in a state bordering on frenzy whenever another bird comes near their habitation. With tail erect and head alert, the father mounts on guard, singing a perfect ecstasy of love to his silent little mate, that sits upon the nest if no danger threatens ; but both rush with passionate malice upon the first intruder, for it must be admitted that Jenny wren is a sad shrew.

While the little family is being reared, or, indeed, at any time, no one is wise enough to estimate the millions of tiny insects from the garden that find their way into the tireless bills of these wrens.

It is often said that the house wren remains at the north all the year, which, though not a fact, is easily accounted for by the coming of the winter wrens just as the others migrate in the autumn, and by their return to Canada when Jenny wren makes up her feather-bed under the eaves in the spring.

Carolina Wren

(Thryothorus ludovicianus) Wren family

Called also : MOCKING WREN

Length—6 inches. Just a trifle smaller than the English sparrow.
Male and Female—Chestnut-brown above. A whitish streak, beginning at base of bill, passes through the eye to the nape of the neck. Throat whitish. Under parts light buff-brown. Wings and tail finely barred with dark.
Range—United States, from Gulf to northern Illinois and southern New England.
Migrations—A common resident except at northern boundary of range, where it is a summer visitor.

This largest of the wrens appears to be the embodiment of the entire family characteristics: it is exceedingly active, nervous, and easily excited, quick-tempered, full of curiosity, peeping into every hole and corner it passes, short of flight as it is of wing, inseparable from its mate till parted by death, and a gushing lyrical songster that only death itself can silence. It also has the wren-like preference for a nest that is roofed over, but not too near the homes of men.

Undergrowths near water, brush heaps, rocky bits of woodland, are favorite resorts. The Carolina wren decidedly objects to being stared at, and likes to dart out of sight in the midst of the underbrush in a twinkling while the opera-glasses are being focussed.

To let off some of his superfluous vivacity, Nature has provided him with two safety-valves : one is his voice, another is his tail. With the latter he gesticulates in a manner so expressive that it seems to be a certain index to what is passing in his busy little brain—drooping it, after the habit of the catbird, when he becomes limp with the emotion of his love-song, or holding it erect as, alert and inquisitive, he peers at the impudent intruder in the thicket below his perch.

But it is his joyous, melodious, bubbling song that is his chief fascination. He has so great a variety of strains that many people have thought that he learned them from other birds, and so have called him what many ornithologists declare that he is not—a mocking wren. And he is one of the few birds that sing at night—not in his sleep or only by moonlight, but even in the total darkness, just before dawn, he gives us the same wide-awake song that entrances us by day.

Winter Wren

(*Troglodytes hiemalis*) Wren family

Length—4 to 4.5 inches. About one-third smaller than the English sparrow. Apparently only half the size.

Male and Female—Cinnamon-brown above, with numerous short, dusky bars. Head and neck without markings. Underneath rusty, dimly and finely barred with dark brown. Tail short.

Range—United States, east and west, and from North Carolina to the Fur Countries.

Migrations—October. April. Summer resident. Commonly a winter resident in the South and Middle States only.

It all too rarely happens that we see this tiny mouse-like wren in summer, unless we come upon him suddenly and overtake him unawares as he creeps shyly over the mossy logs or runs literally "like a flash" under the fern and through the tan-

117

gled underbrush of the deep, cool woods. His presence there is far more likely to be detected by the ear than the eye.

Throughout the nesting season music fairly pours from his tiny throat; it bubbles up like champagne; it gushes forth in a lyrical torrent and overflows into every nook of the forest, that seems entirely pervaded by his song. While music is everywhere, it apparently comes from no particular point, and, search as you may, the tiny singer still eludes, exasperates, and yet entrances.

If by accident you discover him balancing on a swaying twig, never far from the ground, with his comical little tail erect, or more likely pointing towards his head, what a pert, saucy minstrel he is! You are lost in amazement that so much music could come from a throat so tiny.

Comparatively few of his admirers, however, hear the exquisite notes of this little brown wood-sprite, for after the nesting season is over he finds little to call them forth during the bleak, snowy winter months, when in the Middle and Southern States he may properly be called a neighbor. Sharp hunger, rather than natural boldness, drives him near the homes of men, where he appears just as the house wren departs for the South. With a forced confidence in man that is almost pathetic in a bird that loves the forest as he does, he picks up whatever lies about the house or barn in the shape of food—crumbs from the kitchen door, a morsel from the dog's plate, a little seed in the barn-yard, happily rewarded if he can find a spider lurking in some sheltered place to give a flavor to the unrelished grain. Now he becomes almost tame, but we feel it is only because he must be.

The spot that decided preference leads him to, either winter or summer, is beside a bubbling spring. In the moss that grows near it the nest is placed in early summer, nearly always roofed over and entered from the side, in true wren-fashion; and as the young fledglings emerge from the creamy-white eggs, almost the first lesson they receive from their devoted little parents is in the fine art of bathing. Even in winter weather, when the wren has to stand on a rim of ice, he will duck and splash his diminutive body. It is recorded of a certain little individual that he was wont to dive through the icy water on a December day. Evidently the wrens, as a family, are not far removed in the evolutionary scale from true water-birds.

LONG-BILLED MARSH WRENS.
Life-size.

Long-billed Marsh Wren

(Cistothorus palustris) Wren family

Length—4.5 to 5.2 inches. Actually a little smaller than the English sparrow. Apparently half the size.

Male and Female—Brown above, with white line over the eye, and the back irregularly and faintly streaked with white. Wings and tail barred with darker cinnamon-brown. Underneath white. Sides dusky. Tail long and often carried erect. Bill extra long and slender.

Range—United States and southern British America.

Migrations—May. September. Summer resident.

Sometimes when you are gathering cat-tails in the river marshes an alert, nervous little brown bird rises startled from the rushes and tries to elude you as with short, jerky flight it goes deeper and deeper into the marsh, where even the rubber boot may not follow. It closely resembles two other birds found in such a place, the swamp sparrow and the short-billed marsh wren; but you may know by its long, slender bill that it is not the latter, and by the absence of a bright bay crown that it is not the shyest of the sparrows.

These marsh wrens appear to be especially partial to running water; their homes are not very far from brooks and rivers, preferably those that are affected in their rise and flow by the tides. They build in colonies, and might be called inveterate singers, for no single bird is often permitted to finish his bubbling song without half the colony joining in a chorus.

Still another characteristic of this particularly interesting bird is its unique architectural effects produced with coarse grasses woven into globular form and suspended in the reeds. Sometimes adapting its nest to the building material at hand, it weaves it of grasses and twigs, and suspends it from the limb of a bush or tree overhanging the water, where it swings like an oriole's. The entrance to the nest is invariably on the side.

More devoted homebodies than these little wrens are not among the feathered tribe. Once let the hand of man desecrate their nest, even before the tiny speckled eggs are deposited in it, and off go the birds to a more inaccessible place, where they can enjoy their home unmolested. Thus three or four nests may be made in a summer.

Short-billed Marsh Wren

(Cistothorus stellaris) Wren family

Length—4 to 5 inches. Actually about one-third smaller than the
    English sparrow, but apparently only half its size.
Male and Female—Brown above, faintly streaked with white,
    black, and buff. Wings and tail barred with same. Under-
    neath white, with buff and rusty tinges on throat and breast.
    Short bill.
Range—North America, from Manitoba southward in winter to
    Gulf of Mexico. Most common in north temperate latitudes.
Migrations—Early May. Late September.

Where red-winged blackbirds like to congregate in oozy
pastures or near boggy woods, the little short-billed wren may
more often be heard than seen, for he is more shy, if possible,
than his long-billed cousin, and will dive down into the sedges
at your approach, very much as a duck disappears under water.
But if you see him at all, it is usually while swaying to and fro as
he clings to some tall stalk of grass, keeping his balance by the
nervous, jerky tail motions characteristic of all the wrens, and
singing with all his might. Oftentimes his tail reaches backward
almost to his head in a most exaggerated wren-fashion.

Samuels explains the peculiar habit both the long-billed and
the short-billed marsh wrens have of building several nests in
one season, by the theory that they are made to protect the sitting female, for it is noticed that the male bird always lures a
visitor to an empty nest, and if this does not satisfy his curiosity,
to another one, to prove conclusively that he has no family in
prospect.

Wild rice is an ideal nesting place for a colony of these little
marsh wrens. The home is made of sedge grasses, softly lined
with the softer meadow grass or plant-down, and placed in a
tussock of tall grass, or even upon the ground. The entrance is
on the side. But while fond of moist places, both for a home
and feeding ground, it will be noticed that these wrens have no
special fondness for running water, so dear to their long-billed
relatives. Another distinction is that the eggs of this species,
instead of being so densely speckled as to look brown, are pure
white.

BROWN THRASHER.
½ Life size.

Brown Thrasher

(Harporhynchus rufus) Thrasher and Mocking-bird family

Called also: BROWN THRUSH ; GROUND THRUSH ; RED THRUSH ; BROWN MOCKING-BIRD ; FRENCH MOCK-ING-BIRD; MAVIS

Length—11 to 11.5 inches.   Fully an inch longer than the robin.
Male—Rusty red-brown or rufous above; darkest on wings, which
have two short whitish bands.   Underneath white, heavily
streaked (except on throat) with dark-brown, arrow-shaped
spots.   Tail very long.   Yellow eyes.   Bill long and curved
at tip.
Female—Paler than male.
Range—United States to Rockies.   Nests from Gulf States to
Manitoba and Montreal.   Winters south of Virginia.
Migrations—Late April.   October.   Common summer resident.

> " There's a merry brown thrush sitting up in a tree;
> He is singing to me ! He is singing to me !
> And what does he say, little girl, little boy ?
> ' Oh, the world's running over with joy ! ' "

The hackneyed poem beginning with this stanza that delighted our nursery days, has left in our minds a fairly correct
impression of the bird.   He still proves to be one of the perennially joyous singers, like a true cousin of the wrens, and when
we study him afield, he appears to give his whole attention to
his song with a self-consciousness that is rather amusing than
the reverse.   "What musician wouldn't be conscious of his own
powers," he seems to challenge us, "if he possessed such a gift ?"
Seated on a conspicuous perch, as if inviting attention to his performance, with uplifted head and drooping tail he repeats the
one exultant, dashing air to which his repertoire is limited, without waiting for an encore.   Much practice has given the notes a
brilliancy of execution to be compared only with the mocking-
bird's ; but in spite of the name "ferruginous mocking-bird"
that Audubon gave him, he does not seem to have the faculty of
imitating other birds' songs.   Thoreau says the Massachusetts
farmers, when planting their seed, always think they hear the
thrasher say, " Drop it, drop it—cover it up, cover it up—pull it
up, pull it up, pull it up."

One of the shatterings of childish impressions that age too often brings is when we learn by the books that our "merry brown thrush" is no thrush at all, but a thrasher—first cousin to the wrens, in spite of his speckled breast, large size, and certain thrush-like instincts, such as never singing near the nest and shunning mankind in the nesting season, to mention only two. Certainly his bold, swinging flight and habit of hopping and running over the ground would seem to indicate that he is not very far removed from the true thrushes. But he has one undeniable wren-like trait, that of twitching, wagging, and thrashing his long tail about to help express his emotions. It swings like a pendulum as he rests on a branch, and thrashes about in a most ludicrous way as he is feeding on the ground upon the worms, insects, and fruit that constitute his diet.

Before the fatal multiplication of cats, and in unfrequented, sandy locations still, the thrasher builds her nest upon the ground, thus earning the name "ground thrush" that is often given her ; but with dearly paid-for wisdom she now most frequently selects a low shrub or tree to cradle the two broods that all too early in the summer effectually silence the father's delightful song.

Wilson's Thrush

(Turdus fuscescens) Thrush family

Called also : VEERY; TAWNY THRUSH

Length—7 to 7.5 inches. About one-fourth smaller than the robin.

Male and Female—Uniform olive-brown, with a tawny cast above. Centre of the throat white, with cream-buff on sides of throat and upper part of breast, which is lightly spotted with wedge-shaped, brown points. Underneath white, or with a faint grayish tinge.

Range—United States, westward to plains.

Migrations—May. October. Summer resident.

To many of us the veery, as they call the Wilson's thrush in New England, is merely a voice, a sylvan mystery, reflecting the sweetness and wildness of the forest, a vocal "will-o'-the-wisp" that, after enticing us deeper and deeper into the woods, where we sink into the spongy moss of its damp retreats and become

WOOD THRUSH
Life-size.

entangled in the wild grape-vines twined about the saplings and underbrush, still sings to us from unapproachable tangles. Plainly, if we want to see the bird, we must let it seek us out on the fallen log where we have sunk exhausted in the chase.

Presently a brown bird scuds through the fern. It is a thrush, you guess in a minute, from its slender, graceful body. At first you notice no speckles on its breast, but as it comes nearer, obscure arrow-heads are visible—not heavy, heart-shaped spots such as plentifully speckle the larger wood thrush or the smaller hermit. It is the smallest of the three commoner thrushes, and it lacks the ring about the eye that both the others have. Shy and elusive, it slips away again in a most unfriendly fashion, and is lost in the wet tangle before you have become acquainted. You determine, however, before you leave the log, to cultivate the acquaintance of this bird the next spring, when, before it mates and retreats to the forest, it comes boldly into the gardens and scratches about in the dry leaves on the ground for the lurking insects beneath. Miss Florence Merriam tells of having drawn a number of veeries about her by imitating their call-note, which is a whistled *wheew, whoit,* very easy to counterfeit when once heard. "*Taweel-ah, taweel-ah, twil-ah, twil-ah!*" Professor Ridgeway interprets their song, that descends in a succession of trills without break or pause ; but no words can possibly convey an idea of the quality of the music. The veery, that never claims an audience, sings at night also, and its weird, sweet strains floating through the woods at dusk, thrill one like the mysterious voice of a disembodied spirit.

Whittier mentions the veery in " The Playmate " :

> " And here in spring the veeries sing
> The song of long ago."

Wood Thrush

(Turdus mustelinus) Thrush family

Called also : SONG THRUSH ; WOOD ROBIN ; BELLBIRD

Length—8 to 8.3 inches. About two inches shorter than the robin.

Male and Female—Brown above, reddish on head and shoulders, and shading into olive-brown on tail. Throat, breast, and underneath white, plain in the middle, but heavily marked on sides and breast with heart-shaped spots of very dark brown. Whitish eye-ring.

Migrations—Late April or early May. October. Summer resident

When Nuttall wrote of "this solitary and retiring songster," before the country was as thickly settled as it is to-day, it possibly had not developed the confidence in men that now distinguishes the wood thrush from its shy congeners that are distinctly wood birds, which it can no longer strictly be said to be. In city parks and country places, where plenty of trees shade the village streets and lawns, it comes near you, half hopping, half running, with dignified unconsciousness and even familiarity, all the more delightful in a bird whose family instincts should take it into secluded woodlands with their shady dells. Perhaps, in its heart of hearts, it still prefers such retreats. Many conservative wood thrushes keep to their wild haunts, and it must be owned not a few liberals, that discard family traditions at other times, seek the forest at nesting time. But social as the wood thrush is and abundant, too, it is also eminently high-bred; and when contrasted with its tawny cousin, the veery, that skulks away to hide in the nearest bushes as you approach, or with the hermit thrush, that pours out its heavenly song in the solitude of the forest, how gracious and full of gentle confidence it seems! Every gesture is graceful and elegant; even a wriggling beetle is eaten as daintily as caviare at the king's table. It is only when its confidence in you is abused, and you pass too near the nest, that might easily be mistaken for a robin's, just above your head in a sapling, that the wood thrush so far forgets itself as to become excited. *Pit, pit, pit,* sharply reiterated, is called out at you with a strident quality in the tone that is painful evidence of the fearful anxiety your presence gives this gentle bird.

Too many guardians of nests, whether out of excessive happiness or excessive stupidity, have a dangerous habit of singing very near them. Not so the wood thrush. "Come to me," as the opening notes of its flute-like song have been freely translated, invites the intruder far away from where the blue eggs lie cradled in ambush. "*Uoli-a-e-o-li-noli-nol-aeolee-lee!*" is as good a rendering into syllables of the luscious song as could very well be made. Pure, liquid, rich, and luscious, it rings out from the trees on the summer air and penetrates our home like a strain of music from a stringed quartette.

HERMIT THRUSH.
¾ Life-size.

Hermit Thrush

(Turdus aonalaschkæ pallasii) Thrush family

Called also : SWAMP ANGEL; LITTLE THRUSH

Length—7.25 to 7.5 inches. About one-fourth smaller than the robin.

Male and Female—Upper parts olive-brown, reddening near the tail, which is pale rufous, quite distinct from the color of the back. Throat, sides of neck, and breast pale buff. Feathers of throat and neck finished with dark arrow-points at tip; feathers of the breast have larger rounded spots. Sides brownish gray. Underneath white. A yellow ring around the eye. Smallest of the thrushes.

Range—Eastern parts of North America. Most common in the United States to the plains. Winters from southern Illinois and New Jersey to Gulf.

Migrations—April. November. Summer resident.

The first thrush to come and the last to go, nevertheless the hermit is little seen throughout its long visit north. It may loiter awhile in the shrubby roadsides, in the garden or the parks in the spring before it begins the serious business of life in a nest of moss, coarse grass, and pine-needles placed on the ground in the depths of the forest, but by the middle of May its presence in the neighborhood of our homes becomes only a memory. Although one never hears it at its best during the migrations, how one loves to recall the serene, ethereal evening hymn! "The finest sound in Nature," John Burroughs calls it. " It is not a proud, gorgeous strain like the tanager's or the grosbeak's," he says; "it suggests no passion or emotion—nothing personal, but seems to be the voice of that calm, sweet solemnity one attains to in his best moments. It realizes a peace and a deep, solemn joy that only the finest souls may know."

Beyond the question of even the hypercritical, the hermit thrush has a more exquisitely beautiful voice than any other American bird, and only the nightingale's of Europe can be compared with it. It is the one theme that exhausts all the ornithologists' musical adjectives in a vain attempt to convey in words any idea of it to one who has never heard it, for the quality of the song is as elusive as the bird itself. But why should the poets be so silent? Why has it not called forth such verse as the

English poets have lavished upon the nightingale? Undoubtedly because it lifts up its heavenly voice in the solitude of the forest, whereas the nightingales, singing in loud choruses in the moon-light under the poet's very window, cannot but impress his waking thoughts and even his dreams with their melody.

Since the severe storm and cold in the Gulf States a few win-ters ago, where vast numbers of hermit thrushes died from cold and starvation, this bird has been very rare in haunts where it used to be abundant. The other thrushes escaped because they spend the winter farther south.

Alice's Thrush

(Turdus aliciæ) Thrush family

Called also : GRAY-CHEEKED THRUSH

Length—7.5 to 8 inches. About the size of the bluebird.

Male and Female—Upper parts uniform olive-brown. Eye-ring whitish. Cheeks gray; sides dull grayish white. Sides of the throat and breast pale cream-buff, speckled with arrow-shaped points on throat and with half-round dark-brown marks below.

Range—North America, from Labrador and Alaska to Central America.

Migrations—Late April or May. October. Chiefly seen in migra-tions, except at northern parts of its range.

One looks for a prettier bird than this least attractive of all the thrushes in one that bears such a suggestive name. Like the olive-backed thrush, from which it is almost impossible to tell it when both are alive and hopping about the shrubbery, its plu-mage above is a dull olive-brown that is more protective than pleasing.

Just as Wilson hopelessly confused the olive-backed thrush with the hermit, so has Alice's thrush been confounded by later writers with the olive-backed, from which it differs chiefly in being a trifle larger, in having gray cheeks instead of buff, and in possessing a few faint streaks on the throat. Where it goes to make a home for its greenish-blue speckled eggs in some low bush at the northern end of its range, it bursts into song, but except in the nesting grounds its voice is never heard. Mr.

Bradford Torrey, who heard it singing in the White Mountains, describes the song as like the thrush's in quality, but differently accented: *" Wee-o-wee-o-tit-ti-wee-o !"*

In New England and New York this thrush is most often seen during its autumn migrations. As it starts up and perches upon a low branch before you, it appears to have longer legs and a broader, squarer tail than its congeners.

Olive-backed Thrush

(Turdus ustulatus swainsonii) Thrush family

Called also: SWAINSON'S THRUSH

Length—7 to 7.50 inches. About one-fourth smaller than the robin.

Male and Female—Upper parts olive-brown. Whole throat and breast yellow-buff, shading to ashy on sides and to white underneath. Buff ring around eye. Dark streaks on sides of throat (none on centre), and larger, more spot-like marks on breast.

Range—North America to Rockies ; a few stragglers on Pacific slope. Northward to arctic countries.

Migrations—April. October. Summer resident in Canada. Chiefly a migrant in United States.

Mr. Parkhurst tells of finding this "the commonest bird in the Park (Central Park, New York), not even excepting the robin," during the last week of May on a certain year ; but usually, it must be owned, we have to be on the lookout to find it, or it will pass unnoticed in the great companies of more conspicuous birds travelling at the same time. White-throated sparrows often keep it company on the long journeys northward, and they may frequently be seen together, hopping sociably about the garden, the thrush calling out a rather harsh note—*puk ! puk !*—quite different from the liquid, mellow calls of the other thrushes, to resent either the sparrows' bad manners or the inquisitiveness of a human disturber of its peace. But this gregarious habit and neighborly visit end even before acquaintance fairly begins, and the thrushes are off for their nesting grounds in the pine woods of New England or Labrador if they are travelling up the east coast, or to Alaska, British Columbia, or Manitoba if west of the

Mississippi. There they stay all summer, often travelling south-ward with the sparrows in the autumn, as in the spring.

Why they should prefer coniferous trees, unless to utilize the needles for a nest, is not understood. Low trees and bushes are favorite building sites with them as with others of the family, though these thrushes disdain a mud lining to their nests. Those who have heard the olive-backed thrush singing an even-song to its brooding mate compare it with the veery's, but it has a break in it and is less simple and pleasing than the latter's.

Louisiana Water Thrush

(Seiurus motacilla) Wood Warbler family

Length—6 to 6.28 inches. Just a trifle smaller than the English sparrow.

Male and Female—Grayish olive-brown upper parts, with con-spicuous white line over the eye and reaching almost to the nape. Underneath white, tinged with pale buff. Throat and line through the middle, plain. Other parts streaked with very dark brown, rather faintly on the breast, giving them the speckled breast of the thrushes. Heavy, dark bill.

Range—United States, westward to the plains; northward to southern New England. Winters in the tropics.

Migrations—Late April. October. Summer resident.

This bird, that so delighted Audubon with its high-trilled song as he tramped with indefatigable zeal through the hammocks of the Gulf States, seems to be almost the counterpart of the Northern water thrush, just as the loggerhead is the Southern counterpart of the Northern shrike. Very many Eastern birds have their duplicates in Western species, as we all know, and it is most interesting to trace the slight external variations that differ-ent climates and diet have produced on the same bird, and thus differentiated the species. In winter the Northern water thrush visits the cradle of its kind, the swamps of Louisiana and Florida, and, no doubt, by daily contact with its congeners there, keeps close to their cherished traditions, from which it never deviates farther than Nature compels, though it penetrate to the arctic regions during its summer journeys.

With a more southerly range, the Louisiana water thrush does not venture beyond the White Mountains and to the shores of the Great Lakes in summer, but even at the North the same woods often contain both birds, and there is opportunity to note just how much they differ. The Southern bird is slightly the larger, possibly an inch; it is more gray, and it lacks a few of the streaks, notably on the throat, that plentifully speckle its Northern counterpart; but the habits of both of these birds appear to be identical. Only for a few days in the spring or autumn migrations do they pass near enough to our homes for us to study them, and then we must ever be on the alert to steal a glance at them through the opera-glasses, for birds more shy than they do not visit the garden shrubbery at any season. Only let them suspect they are being stared at, and they are under cover in a twinkling.

Where mountain streams dash through tracts of mossy, spongy ground that is carpeted with fern and moss, and overgrown with impenetrable thickets of underbrush and tangles of creepers—such a place is the favorite resort of both the water thrushes. With a rubber boot missing, clothes torn, and temper by no means unruffled, you finally stand over the Louisiana thrush's nest in the roots of an upturned tree immediately over the water, or else in a mossy root-belaced bank above a purling stream. A liquid-trilled warble, wild and sweet, breaks the stillness, and, like Audubon, you feel amply rewarded for your pains, though you may not be prepared to agree with him in thinking the song the equal of the European nightingale's.

Northern Water Thrush

(Seiurus noveboracensis) Wood Warbler family

Called also : NEW YORK WATER THRUSH ; AQUATIC WOOD WAGTAIL ; AQUATIC THRUSH

Length—5 to 6 inches. A trifle smaller than the English sparrow.
Male and Female—Uniform olive or grayish brown above. Pale buff line over the eye. Underneath, white tinged with sulphur-yellow, and streaked like a thrush with very dark brown arrow-headed or oblong spots that are also seen underneath wings.
Range—United States, westward to Rockies and northward through British provinces. Winters from Gulf States southward.
Migrations—Late April. October. Summer resident.

According to the books we have before us, a warbler; but who, to look at his speckled throat and breast, would ever take him for anything but a diminutive thrush; or, studying him from some distance through the opera-glasses as he runs in and out of the little waves along the brook or river shore, would not name him a baby sandpiper? The rather unsteady motion of his legs, balancing of the tail, and sudden jerking of the head suggest an aquatic bird rather than a bird of the woods. But to really know either man or beast, you must follow him to his home, and if you have pluck enough to brave the swamp and the almost impenetrable tangle of undergrowth where the water thrush chooses to nest, there "In the swamp in secluded recesses, a shy and hidden bird is warbling a song;" and this warbled song that Walt Whitman so adored gives you your first clue to the proper classification of the bird. It has nothing in common with the serene, hymn-like voices of the true thrushes ; the bird has no flute-like notes, but an emphatic smacking or chucking kind of warble. For a few days only is this song heard about the gardens and roadsides of our country places. Like the Louisiana water thrush, this bird never ventures near the homes of men after the spring and autumn migrations, but, on the contrary, goes as far away from them as possible, preferably to some mountain region, beside a cool and dashing brook, where a party of adventurous young climbers from a summer hotel or the lonely trout fisherman may startle it from its mossy nest on the ground.

Flicker

(Colaptes auratus) Woodpecker family

Called also: GOLDEN-WINGED WOODPECKER ; CLAPE ; PIGEON WOODPECKER ; YELLOWHAMMER ; HIGH-HOLE OR HIGH-HOLDER ; YARUP ; WAKE-UP ; YELLOW-SHAFTED WOODPECKER

Length—12 to 13 inches. About one-fourth as large again as the robin.

Male and Female—Head and neck bluish gray, with a red crescent across back of neck and a black crescent on breast. Male has black cheek-patches, that are wanting in female. Golden brown shading into brownish-gray, and barred with black above. Underneath whitish, tinged with light chocolate

FLICKER.
½ Life-size.

and thickly spotted with black. Wing linings, shafts of wing, and tail-quills bright yellow. Above tail white, conspicuous when the bird flies.

Range—United States, east of Rockies; Alaska and British America, south of Hudson Bay. Occasional on Pacific slope.

Migrations—Most commonly seen from April to October. Usually resident.

If we were to follow the list of thirty-six aliases by which this largest and commonest of our woodpeckers is known throughout its wide range, we should find all its peculiarities of color, flight, noises, and habits indicated in its popular names. It cannot but attract attention wherever seen, with its beautiful plumage, conspicuously yellow if its outstretched wings are looked at from below, conspicuously brown and white if seen upon the ground. At a distance it suggests the meadowlark. Both birds wear black, crescent breast decorations, and the flicker also has the habit of feeding upon the ground, especially in autumn, a characteristic not shared by its relations.

Early in the spring this bird of many names and many voices makes itself known by a long, strong, sonorous call, a sort of proclamation that differs from its song proper, which Audubon calls "a prolonged jovial laugh" (described by Mrs. Wright as "*Wick, wick, wick, wick!*"), and differs also from its rapidly repeated, mellow, and most musical *cuh, cuh, cuh, cuh, cuh*, uttered during the nesting season.

Its nasal *kee-yer*, vigorously called out in the autumn, is less characteristic, however, than the sound it makes while associating with its fellows on the feeding ground—a sound that Mr. Frank M. Chapman says can be closely imitated by the swishing of a willow wand.

A very ardent and ridiculous-looking lover is this bird, as, with tail stiffly spread, he sidles up to his desired mate and bows and bobs before her, then retreats and advances, bowing and bobbing again, very often with a rival lover beside him (whom he generously tolerates) trying to outdo him in grace and general attractiveness. Not the least of the bird's qualities that must commend themselves to the bride is his unfailing good nature, genial alike in the home and in the field.

The "high-holders" have the peculiar and silly habit of boring out a number of superfluous holes for nests high up in the trees, in buildings, or hollow wooden columns, only one of which they intend to use. Six white eggs is the proper number for a household, but Dr. Coues says the female that has been robbed keeps on laying three or even four sets of eggs without interruption.

Meadowlark

(Sturnella magna) Blackbird family

Called also : FIELD LARK ; OLDFIELD LARK

Length—10 to 11 inches. A trifle larger than the robin.

Male—Upper parts brown, varied with chestnut, deep brown, and black. Crown streaked with brown and black, and with a cream-colored streak through the centre. Dark-brown line apparently running through the eye ; another line over the eye, yellow. Throat and chin yellow ; a large, conspicuous black crescent on breast. Underneath yellow, shading into buffy brown, spotted or streaked with very dark brown. Outer tail feathers chiefly white, conspicuous in flight. Long, strong legs and claws, adapted for walking. Less black in winter plumage, which is more grayish brown.

Female—Paler than male.

Range—North America, from Newfoundland to the Gulf of Mexico, and westward to the plains, where the Western meadowlark takes its place. Winters from Massachusetts and Illinois southward.

Migrations—April. Late October. Usually a resident, a few remaining through the winter.

In the same meadows with the red-winged blackbirds, birds of another feather, but of the same family, nevertheless, may be found flocking together, hunting for worms and larvæ, building their nests, and rearing their young very near each other with the truly social instinct of all their kin.

The meadowlarks, which are really not larks at all, but the blackbirds' and orioles' cousins, are so protected by the coloring of the feathers on their backs, like that of the grass and stubble they live among, that ten blackbirds are noticed for every meadowlark, although the latter is very common. Not until you flush a flock of them as you walk along the roadside or through the meadows and you note the white tail feathers and the black crescents on the yellow breasts of the large brown birds that rise

towards the tree-tops with whirring sound and a flight suggesting the quail's, do you suspect there are any birds among the tall grasses.

Their clear and piercing whistle, "*Spring o' the y-e-a r. Spring o' the year!*" rings out from the trees with varying intonation and accent, but always sweet and inspiriting. To the bird's high vantage ground you may not follow, for no longer having the protection of the high grass, it has become wary and flies away as you approach, calling out *peent-peent* and nervously flitting its tail (again showing the white feather), when it rests a moment on the pasture fence-rail.

It is like looking for a needle in a haystack to try to find a meadowlark's nest, an unpretentious structure of dried grasses partly arched over and hidden in a clump of high timothy, flat upon the ground. But what havoc snakes and field-mice play with the white-speckled eggs and helpless fledglings ! The care of rearing two or three broods in a season and the change of plumage to duller winter tints seem to exhaust the high spirits of the sweet whistler. For a time he is silent, but partly regains his vocal powers in the autumn, when, with large flocks of his own kind, he resorts to marshy feeding grounds. In the winter he chooses for companions the horned larks, that walk along the shore, or the snow buntings and sparrows of the inland pastures, and will even include the denizens of the barn-yard when hunger drives him close to the haunts of men.

The Western Meadowlark or Prairie Lark (*Sturnella magna neglecta*), which many ornithologists consider a different species from the foregoing, is distinguished chiefly by its lighter, more grayish-brown plumage, by its yellow cheeks, and more especially by its richer, fuller song. In his "Birds of Manitoba" Mr. Ernest E. Thompson says of this meadowlark : "In richness of voice and modulation it equals or excels both wood thrush and nightingale, and in the beauty of its articulation it has no superior in the whole world of feathered choristers with which I am acquainted."

Horned Lark

(Otocoris alpestris) Lark family

Called also: SHORE LARK

Length—7.5 to 8 inches. About one-fifth smaller than the robin.

Male—Upper parts dull brown, streaked with lighter on edges and tinged with pink or vinaceous; darkest on back of head, neck, shoulders, and nearest the tail. A few erectile feathers on either side of the head form slight tufts or horns that are wanting in female. A black mark from the base of the bill passes below the eye and ends in a horn-shaped curve on cheeks, which are yellow. Throat clear yellow. Breast has crescent-shaped black patch. Underneath soiled white, with dusky spots on lower breast. Tail black, the outer feathers margined with white, noticed in flight.

Female—Has yellow eye-stripe; less prominent markings, especially on head, and is a trifle smaller.

Range—Northeastern parts of North America, and in winter from Ohio and eastern United States as far south as North Carolina.

Migrations—October and November. March. Winter resident.

Far away to the north in Greenland and Labrador this true lark, the most beautiful of its genus, makes its summer home. There it is a conspicuously handsome bird with its pinkish-gray and chocolate feathers, that have greatly faded into dull browns when we see them in the late autumn. In the far north only does it sing, and, according to Audubon, the charming song is flung to the breeze while the bird soars like a skylark. In the United States we hear only its call-note.

Great flocks come down the Atlantic coast in October and November, and separate into smaller bands that take up their residence in sandy stretches and open tracts near the sea or wherever the food supply looks promising, and there the larks stay until all the seeds, buds of bushes, berries, larvæ, and insects in their chosen territory are exhausted. They are ever conspicuously ground birds, walkers, and when disturbed at their dinner, prefer to squat on the earth rather than expose themselves by flight. Sometimes they run nimbly over the frozen ground to escape an intruder, but flying they reserve as a last resort. When the visitor has passed they quickly return to their dinner. If they were content to eat less ravenously and remain slender, fewer

HORNED LARK.

¼ Life-size.

victims might be slaughtered annually to tickle the palates of the epicure. It is a mystery what they find to fatten upon when snow covers the frozen ground. Even in the severe midwinter storms they will not seek the protection of the woods, but always prefer sandy dunes with their scrubby undergrowth or open meadow lands. Occasionally a small flock wanders toward the farms to pick up seeds that are blown from the hayricks or scattered about the barn-yard by overfed domestic fowls.

The Prairie Horned Lark *(Otocoris alpestris praticola)* is similar to the preceding, but a trifle smaller and paler, with a white instead of a yellow streak above the eye, the throat yellowish or entirely white instead of sulphur-yellow, and other minor differences. It has a far more southerly range, confined to northern portions of the United States from the Mississippi eastward. Once a distinctly prairie bird, it now roams wherever large stretches of open country that suit its purposes are cleared in the East, and remains resident. This species also sings in midair on the wing, but its song is a crude, half-inarticulate affair, barely audible from a height of two hundred feet.

American Pipit

(Anthus pensilvanicus) Wagtail family

Called also: TITLARK; BROWN OR RED LARK

Length—6.38 to 7 inches. About the size of a sparrow.

Male and Female—Upper parts brown; wings and tail dark olive-brown; the wing coverts tipped with buff or whitish, and ends of outer tail feathers white, conspicuous in flight. White or yellowish eye-ring, and line above the eye. Underneath light buff brown, with spots on breast and sides, the under parts being washed with brown of various shades. Feet brown. Hind toe-nail as long as or longer than the toe.

Range—North America at large. Winters south of Virginia to Mexico and beyond.

Migrations—April. October or November. Common in the United States, chiefly during the migrations.

The color of this bird varies slightly with age and sex, the under parts ranging from white through pale rosy brown to a reddish tinge; but at any season, and under all circumstances, the pipit is a distinctly brown bird, resembling the water thrushes not in plumage only, but in the comical tail waggings and jerkings that alone are sufficient to identify it. However the books may tell us the bird is a wagtail, it certainly possesses two strong characteristics of true larks: it is a walker, delighting in walking or running, never hopping over the ground, and it has the angelic habit of singing as it flies.

During the migrations the pipits are abundant in salt marshes or open stretches of country inland, that, with lark-like preference, they choose for feeding grounds. When flushed, all the flock rise together with uncertain flight, hovering and wheeling about the place, calling down *dee-dee, dee-dee* above your head until you have passed on your way, then promptly returning to the spot from whence they were disturbed. Along the roadsides and pastures, where two or three birds are frequently seen together, they are too often mistaken for the vesper sparrows because of their similar size and coloring, but their easy, graceful walk should distinguish them at once from the hopping sparrow. They often run to get ahead of some one in the lane, but rarely fly if they can help it, and then scarcely higher than a fence-rail. Early in summer they are off for the mountains in the north. Labrador is their chosen nesting ground, and they are said to place their grassy nest, lined with lichens or moss, flat upon the ground—still another lark trait. Their eggs are chocolate-brown scratched with black.

Whippoorwill

(Antrostomus vociferus) Goatsucker family

Length—9 to 10 inches. About the size of the robin. Apparently much larger, because of its long wings and wide wing-spread.

Male—A long-winged bird, mottled all over with reddish brown, grayish black, and dusky white; numerous bristles fringing the large mouth. A narrow white band across the upper breast. Tail quills on the end and under side white.

Female—Similar to male, except that the tail is dusky in color where that of the male is white. Band on breast buff instead of white.

Range—United States, to the plains. Not common near the sea.

Migrations—Late April to middle of September. Summer resident.

The whippoorwill, because of its nocturnal habits and plaintive note, is invested with a reputation for occult power which inspires a chilling awe among superstitious people, and leads them insanely to attribute to it an evil influence ; but it is a harmless, useful night prowler, flying low and catching enormous numbers of hurtful insects, always the winged varieties, in its peculiar fly-trap mouth.

It loves the rocky, solitary woods, where it sleeps all day; but it is seldom seen, even after painstaking search, because of its dull, mottled markings conforming so nearly to rocks and dry leaves, and because of its unusual habit of stretching itself lengthwise on a tree branch or ledge, where it is easily confounded with a patch of lichen, and thus overlooked. If by accident one happens upon a sleeping bird, it suddenly rouses and flies away, making no more sound than a passing butterfly—a curious and uncanny silence that is quite remarkable. When the sun goes down and as the gloaming deepens, the bird's activity increases, and it begins its nightly duties, emitting from time to time, like a sentry on his post or a watchman of the night, the doleful call which has given the bird its common name. It

" Mourns unseen, and ceaseless sings
Ever a note of wail and woe,"

that our Dutch ancestors interpreted as "*Quote-kerr-kee*," and so called it. They had a tradition that no frost ever appeared after the bird had been heard calling in the spring, and that it wisely left for warmer skies before frost came in the autumn. Prudent bird, never caught napping !

It is erratic in its choice of habitations, even when rock and solitude seem suited to its taste. Under no stress of circumstances is it found close to the seashore, and in the Hudson River valley it keeps a half mile or more back from the river.

The eggs, generally two in number, are creamy white, dashed with dark and olive spots, and laid on the ground on dry leaves, or in a little hollow in rock or stump—never in a nest built with loving care. But in extenuation of such carelessness it may be said that, if disturbed or threatened, the mother shows no lack of maternal instinct, and removes her young, carrying them in her beak as a cat conveys her kittens to secure shelter.

Nighthawk

(Chordeiles virginianus) Goatsucker family

Called also : NIGHTJAR; BULL-BAT; MOSQUITO HAWK;
WILL - O' - THE - WISP ; PISK ; PIRAMIDIG ; LONG -
WINGED GOATSUCKER

Length—9 to 10 inches. About the same length as the robin, but
   apparently much longer because of its very wide wing-spread.

Male and Female—Mottled blackish brown and rufous above, with
   a multitude of cream-yellow spots and dashes. Lighter
   below, with waving bars of brown on breast and underneath. White mark on throat, like an imperfect horseshoe;
   also a band of white across tail of male bird. These latter
   markings are wanting in female. Heavy wings, which are
   partly mottled, are brown on shoulders and tips, and longer
   than tail. They have large white spots, conspicuous in
   flight, one of their distinguishing marks from the whippoorwill. Head large and depressed, with large eyes and ear-
   openings. Very small bill.

Range—From Mexico to arctic islands.

Migrations—May. October. Common summer resident.

The nighthawk's misleading name could not well imply
more that the bird is not : it is not nocturnal in its habits, neither
is it a hawk, for if it were, no account of it would be given in
this book, which distinctly excludes birds of prey. Stories of its
chicken-stealing prove to be ignorant rather than malicious slan-
ders. Any one disliking the name, however, surely cannot complain of a limited choice of other names by which, in different
sections of the country, it is quite as commonly known.

Too often it is mistaken for the whippoorwill. The nighthawk does not have the weird and woful cry of that more dismal
bird, but gives instead a harsh, whistling note while on the wing,
followed by a vibrating, booming, whirring sound that Nuttall
likens to "the rapid turning of a spinning wheel, or a strong
blowing into the bung-hole of an empty hogshead." This peculiar sound is responsible for the name nightjar, frequently given
to this curious bird. It is said to be made as the bird drops suddenly through the air, creating a sort of stringed instrument of its
outstretched wings and tail. When these wings are spread, their
large white spots running through the feathers to the under side

138

NIGHT HAWK.
¾ Life-size.

should be noted to further distinguish the nighthawk from the whippoorwill, which has none, but which it otherwise closely resembles. This booming sound, coming from such a height that the bird itself is often unseen, was said by the Indians to be made by the shad spirits to warn the scholes of shad about to ascend the rivers to spawn in the spring, of their impending fate.

The flight of the nighthawk is free and graceful in the extreme. Soaring through space without any apparent motion of its wings, suddenly it darts with amazing swiftness like an erratic bat after the fly, mosquito, beetle, or moth that falls within the range of its truly hawk-like eye.

Usually the nighthawks hunt in little companies in the most sociable fashion. Late in the summer they seem to be almost gregarious. They fly in the early morning or late afternoon with beak wide open, hawking for insects, but except when the moon is full they are not known to go a-hunting after sunset. During the heat of the day and at night they rest on limbs of trees, fence-rails, stone walls, lichen-covered rocks or old logs—wherever Nature has provided suitable mimicry of their plumage to help conceal them.

With this object in mind, they quite as often choose a hollow surface of rock in some waste pasture or the open ground on which to deposit the two speckled-gray eggs that sixteen days later will give birth to their family. But in August, when family cares have ended for the season, it is curious to find this bird of the thickly wooded country readily adapting itself to city life, resting on Mansard roofs, darting into the streets from the house-tops, and wheeling about the electric lights, making a hearty supper of the little, winged insects they attract.

Black-billed Cuckoo

(Coccyzus erythrophthalmus) Cuckoo family

Called also : RAIN CROW

Length—11 to 12 inches. About one-fifth larger than the robin.
Male—Grayish brown above, with bronze tint in feathers. Underneath grayish white ; bill, which is long as head and black, arched and acute. Skin about the eye bright red. Tail long, and with spots on tips of quills that are small and inconspicuous.

Female—Has obscure dusky bars on the tail.
Range—Labrador to Panama; westward to Rocky Mountains.
Migrations—May. September. Summer resident.

> " O cuckoo ! shall I call thee bird ?
> Or but a wandering voice ? "

From the tangled shrubbery on the hillside back of Dove Cottage, Keswick, where Wordsworth and his sister Dorothy listened for the coming of this " darling of the spring " ; in the willows overhanging Shakespeare's Avon ; from the favorite haunts of Chaucer and Spenser, where

> " Runneth meade and springeth blede, "

we hear the cuckoo calling ; but how many on this side of the Atlantic are familiar with its American counterpart ? Here, too, the cuckoo delights in running water and damp, cloudy weather like that of an English spring; it haunts the willows by our riversides, where as yet no " immortal bard " arises to give it fame. It " loud sings " in our shrubbery, too. Indeed, if we cannot study our bird afield, the next best place to become acquainted with it is in the pages of the English poets. But due allowance must be made for differences of temperament. Our cuckoo is scarcely a " merry harbinger " ; his talents, such as they are, certainly are not musical. However, the guttural cluck is not discordant, and the black-billed species, at least, has a soft, mellow voice that seems to indicate an embryonic songster. " *K-k-k-k, kow-kow-ow-how-ow !* " is a familiar sound in many localities, but the large, slim, pigeon-shaped, brownish-olive bird that makes it, securely hidden in the low trees and shrubs that are its haunts, is not often personally known. Catching a glimpse only of the grayish-white under parts from where we stand looking up into the tree at it, it is quite impossible to tell the bird from the yellow-billed species. When, as it flies about, we are able to note the red circles about its eyes, its black bill, and the absence of black tail feathers, with their white " thumb-nail " spots, and see no bright cinnamon feathers on the wings (the yellow-billed specie's distinguishing marks), we can at last claim acquaintance with the black-billed cuckoo. Our two common cuckoos are so nearly alike that they are constantly confused in the popular mind and very often in the writings of ornithologists. At first glance the birds look alike. Their haunts are almost identical ; their habits are the same ; and, as they usually keep well out of sight, it is not surprising if confusion arise.

Neither cuckoo knows how to build a proper home; a bunch of sticks dropped carelessly into the bush, where the hapless babies that emerge from the greenish eggs will not have far to fall when they tumble out of bed, as they must inevitably do, may by courtesy only be called a nest. The cuckoo is said to suck the eggs of other birds ; but, surely, such vice is only the rarest dissipation. Insects of many kinds and "tent caterpillars" chiefly are their chosen food.

Yellow-billed Cuckoo

(Coccyzus americanus) Cuckoo family

Called also : RAIN CROW

Length—11 to 12 inches. About one-fifth longer than the robin.
Male and Female—Grayish brown above, with bronze tint in feathers. Underneath grayish white. Bill, which is as long as head, arched, acute, and more robust than the black-billed species, and with lower mandible yellow. Wings washed with bright cinnamon-brown. Tail has outer quills black, conspicuously marked with white thumb-nail spots. Female larger.
Range—North America, from Mexico to Labrador. Most common in temperate climates. Rare on Pacific slope.
Migrations—Late April. September. Summer resident.

"*Kak, k-kuk, k-kuk, k-kuk!*" like an exaggerated tree-toad's rattle, is a sound that, when first heard, makes you rush out of doors instantly to "name" the bird. Look for him in the depths of the tall shrubbery or low trees, near running water, if there is any in the neighborhood, and if you are more fortunate than most people, you will presently become acquainted with the yellow-billed cuckoo. When seen perching at a little distance, his large, slim body, grayish brown, with olive tints above and whitish below, can scarcely be distinguished from that of the black-billed species. It is not until you get close enough to note the yellow bill, reddish-brown wings, and black tail feathers with their white "thumb-nail" marks, that you know which cuckoo you are watching. In repose the bird looks dazed or stupid, but as it darts about among the trees after insects, noiselessly slipping to another one that promises better results, and hopping along the limbs after performing a series of beautiful evolutions among the branches as it hunts for its favorite "tent caterpillars," it appears what it really is : an unusually active, graceful, intelligent bird.

A solitary wanderer, nevertheless one cuckoo in an apple orchard is worth a hundred robins in ridding it of caterpillars and inch-worms, for it delights in killing many more of these than it can possibly eat. In the autumn it varies its diet with minute fresh-water shellfish from the swamp and lake. Mulberries, that look so like caterpillars the bird possibly likes them on that account, it devours wholesale.

Family cares rest lightly on the cuckoos. The nest of both species is a ramshackle affair—a mere bundle of twigs and sticks without a rim to keep the eggs from rolling from the bush, where they rest, to the ground. Unlike their European relative, they have the decency to rear their own young and not impose this heavy task on others; but the cuckoos on both sides of the Atlantic are most erratic and irregular in their nesting habits. The overworked mother-bird often lays an egg while brooding over its nearly hatched companion, and the two or three half-grown fledglings already in the nest may roll the large greenish eggs out upon the ground, while both parents are off searching for food to quiet their noisy clamorings. Such distracting mismanagement in the nursery is enough to make a homeless wanderer of any father. It is the mother-bird that tumbles to the ground at your approach from sheer fright ; feigns lameness, trails her wings as she tries to entice you away from the nest. The male bird shows far less concern ; a no more devoted father, we fear, than he is a lover. It is said he changes his mate every year.

Altogether, the cuckoo is a very different sort of bird from what our fancy pictured. The little Swiss creatures of wood that fly out of the doors of clocks and call out the bed-hour to sleepy children, are chiefly responsible for the false impressions of our mature years. The American bird does not repeat its name, and its harsh, grating "*kuk, kuk,*" does not remotely suggest the sweet voice of its European relative.

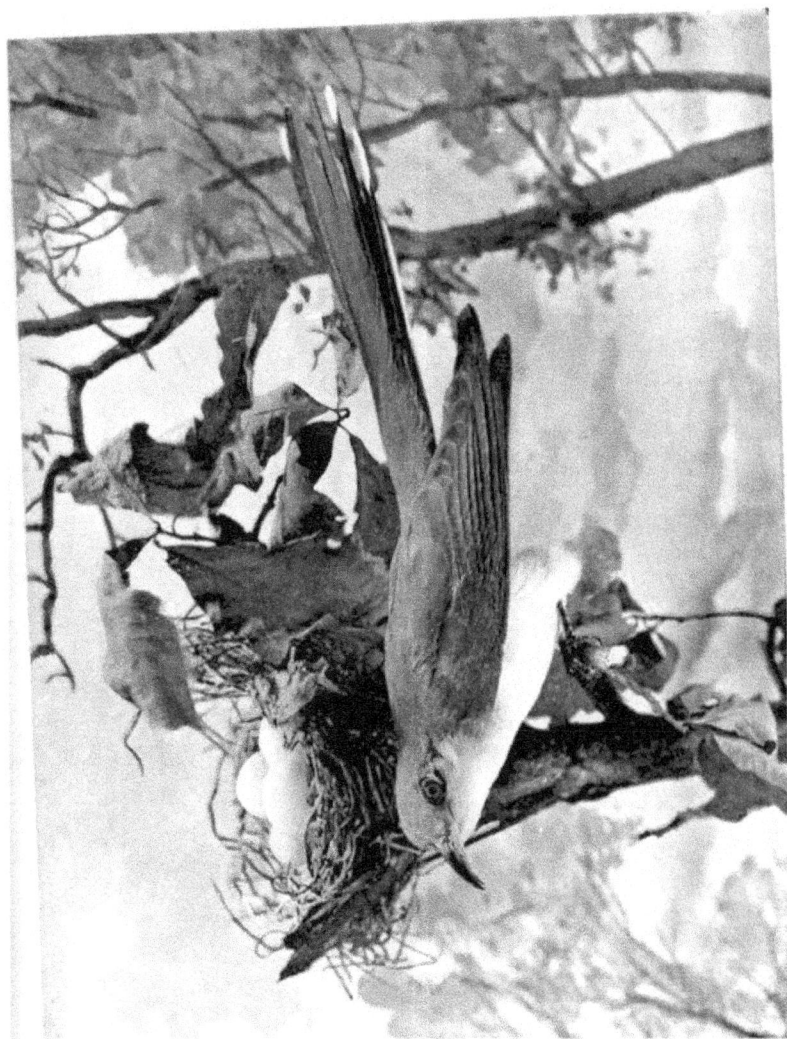

Bank Swallow

(Clivicola riparia) Swallow family

Called also : SAND MARTIN ; SAND SWALLOW

Length—5 to 5.5 inches. About an inch shorter than the English sparrow, but apparently much larger because of its wide wing-spread.

Male and Female—Grayish brown or clay-colored above. Upper wings and tail darkest. Below, white, with brownish band across chest. Tail, which is rounded and more nearly square than the other swallows, is obscurely edged with white.

Range—Throughout North America south of Hudson Bay.

Migrations—April. October. Summer resident.

Where a brook cuts its way through a sand bank to reach the sea is an ideal nesting ground for a colony of sand martins. The face of the high bank shows a number of clean, round holes indiscriminately bored into the sand, as if the place had just received a cannonading; but instead of war an atmosphere of peace pervades the place in midsummer, when you are most likely to visit it. Now that the young ones have flown from their nests that your arm can barely reach through the tunnelled sand or clay, there can be little harm in examining the feathers dropped from gulls, ducks, and other water-birds with which the grassy home is lined.

The bank swallow's nest, like the kingfisher's, which it resembles, is his home as well. There he rests when tired of flying about in pursuit of insect food. Perhaps a bird that has been resting in one of the tunnels, startled by your innocent house-breaking, will fly out across your face, near enough for you to see how unlike the other swallows he is: smaller, plainer, and with none of their glinting steel-blues and buffs about him. With strong, swift flight he rejoins his fellows, wheeling, skimming, darting through the air above you, and uttering his characteristic "giggling twitter," that is one of the cheeriest noises heard along the beach. In early October vast numbers of these swallows may be seen in loose flocks along the Jersey coast, slowly making their way South. Clouds of them miles in extent are recorded.

Closely associated with the sand martin is the Rough-winged Swallow *(Stelgidopteryx serripennis)*, not to be distinguished from its companion on the wing, but easily recognized by its dull-gray throat and the absence of the brown breast-band when seen at close range.

Cedar Bird

(Ampelis cedrorum) Waxwing family

Called also: CEDAR WAXWING; CHERRY-BIRD; CANADA ROBIN; RÉCOLLET

Length—7 to 8 inches. About one-fifth smaller than the robin.

Male—Upper parts rich grayish brown, with plum-colored tints showing through the brown on crest, throat, breast, wings, and tail. A velvety-black line on forehead runs through the eye and back of crest. Chin black; crest conspicuous; breast lighter than the back, and shading into yellow underneath. Wings have quill-shafts of secondaries elongated, and with brilliant vermilion tips like drops of sealing-wax, rarely seen on tail quills, which have yellow bands across the end.

Female—With duller plumage, smaller crest, and narrower tail-band.

Range—North America, from northern British provinces to Central America in winter.

Migrations—A roving resident, without fixed seasons for migrating.

As the cedar birds travel about in great flocks that quickly exhaust their special food in a neighborhood, they necessarily lead a nomadic life—here to-day, gone to-morrow—and, like the Arabs, they "silently steal away." It is surprising how very little noise so great a company of these birds make at any time. That is because they are singularly gentle and refined; soft of voice, as they are of color, their plumage suggesting a fine Japanese water-color painting on silk, with its beautiful sheen and exquisitely blended tints.

One listens in vain for a song; only a lisping "*Twee-twee-ze,*" or "a dreary whisper," as Minot calls their low-toned communications with each other, reaches our ears from their high perches in the cedar trees, where they sit, almost motionless hours at a time, digesting the enormous quantities of juniper and whortle berries, wild cherries, worms, and insects upon which they have gormandized.

Nuttall gives the cedar birds credit for excessive politeness to each other. He says he has often seen them passing a worm from one to another down a whole row of beaks and back again before it was finally eaten.

When nesting time arrives—that is to say, towards the end of the summer—they give up their gregarious habits and live in pairs, billing and kissing like turtle-doves in the orchard or wild crab-trees, where a flat, bulky nest is rather carelessly built of twigs, grasses, feathers, strings—any odds and ends that may be lying about. The eggs are usually four, white tinged with purple and spotted with black.

Apparently they have no moulting season; their plumage is always the same, beautifully neat and full-feathered. Nothing ever hurries or flusters them, their greatest concern apparently being, when they alight, to settle themselves comfortably between their over-polite friends, who are never guilty of jolting or crowding. Few birds care to take life so easily, not to say indolently.

Among the French Canadians they are called Récollet, from the color of their crest resembling the hood of the religious order of that name. Every region the birds pass through, local names appear to be applied to them, a few of the most common of which are given above.

Of the three waxwings known to scientists, two are found in America, and the third in Japan.

Brown Creeper

(Certhia familiaris americana) Creeper family

Length—5 to 5.75 inches. A little smaller than the English sparrow.

Male and Female—Brown above, varied with ashy-gray stripes and small, lozenge-shaped gray mottles. Color lightest on head, increasing in shade to reddish brown near tail. Tail paler brown and long; wings brown and barred with whitish. Beneath grayish white. Slender, curving bill.

Range—United States and Canada, east of Rocky Mountains.

Migrations—April. September. Winter resident.

This little brown wood sprite, the very embodiment of virtuous diligence, is never found far from the nuthatches, titmice, and kinglets, though not strictly in their company, for he is a rather solitary bird. Possibly he repels them by being too exasperatingly conscientious.

Beginning at the bottom of a rough-barked tree (for a smooth bark conceals no larvæ), the creeper silently climbs upward in a sort of spiral, now lost to sight on the opposite side of the tree, then reappearing just where he is expected to, flitting back a foot or two, perhaps, lest he overlooked a single spider egg, but never by any chance leaving a tree until conscience approves of his thoroughness. And yet with all this painstaking workman's care, it takes him just about fifty seconds to finish a tree. Then off he flits to the base of another, to repeat the spiral process. Only rarely does he adopt the woodpecker process of partly flitting, partly rocking his way with the help of his tail straight up one side of the tree.

Yet this little bird is not altogether the soulless drudge he appears. In the midst of his work, uncheered by summer sunshine, and clinging with numb toes to the tree-trunk some bitter cold day, he still finds some tender emotion within him to voice in a "wild, sweet song" that is positively enchanting at such a time. But it is not often this song is heard south of his nesting grounds.

The brown creeper's plumage is one of Nature's most successful feats of mimicry—an exact counterfeit in feathers of the brown-gray bark on which the bird lives. And the protective coloring is carried out in the nest carefully tucked under a piece of loosened bark in the very heart of the tree.

Pine Siskin

(*Spinus pinus*) Finch family

Called also : PINE FINCH ; PINE LINNET

Length—4.75 to 5 inches. Over an inch smaller than the English sparrow.

Male and Female—Olive-brown and gray above, much streaked and striped with very dark brown everywhere. Darkest on head and back. Lower back, base of tail, and wing feathers pale sulphur-yellow. Under parts very light buff brown, heavily streaked.

146

Range—North America generally. Most common in north latitudes. Winters south to the Gulf of Mexico.

Migrations—Erratic winter visitor from October to April. Uncommon in summer.

A small grayish-brown brindle bird, relieved with touches of yellow on its back, wings, and tail, may be seen some winter morning roving on the lawn from one evergreen tree to another, clinging to the pine cones and peering attentively between the scales before extracting the kernels. It utters a call-note so like the English sparrow's that you are surprised when you look up into the tree to find it comes from a stranger. The pine siskin is an erratic visitor, and there is always the charm of the unexpected about its coming near our houses that heightens our enjoyment of its brief stay.

As it flies downward from the top of the spruce tree to feed upon the brown seeds still clinging to the pigweed and goldenrod stalks sticking out above the snow by the roadside, it dips and floats through the air like its charming little cousin, the goldfinch. They have several characteristics in common besides their flight and their fondness for thistles. Far at the north, where the pine siskin nests in the top of the evergreens, his sweet-warbled love-song is said to be like that of our "wild canary's," only with a suggestion of fretfulness in the tone.

Occasionally some one living in an Adirondack or other mountain camp reports finding the nest and hearing the siskin sing even in midsummer; but it is, nevertheless, considered a northern species, however its erratic habits may sometimes break through the ornithologist's traditions.

Smith's Painted Longspur

(Calcarius pictus) Finch family

Length—6.5 inches. About the size of a large English sparrow.

Male and Female—Upper parts marked with black, brown, and white, like a sparrow; brown predominant. Male bird with more black about head, shoulders, and tail feathers, and a whitish patch, edged with black, under the eye. Underneath pale brown, shading to buff. Hind claw or spur conspicuous.

147

Range—Interior of North America, from the arctic coast to Illinois
    and Texas.
Migrations—Winter visitor.   Without fixed season.

Confined to a narrower range than the Lapland longspur,
this bird, quite commonly found on the open prairie districts of
the middle West in winter, is, nevertheless, so very like its cousin
that the same description of their habits might very well answer
for both. Indeed, both these birds are often seen in the same
flock. Larks and the ubiquitous sparrows, too, intermingle with
them with the familiarity that only the starvation rations of midwinter, and not true sociability, can effect ; and, looking out upon
such a heterogeneous flock of brown birds as they are feeding
together on the frozen ground, only the trained field ornithologist
would find it easy to point out the painted longspurs.

Certain peculiarities are noticeable, however. Longspurs
squat while resting ; then, when flushed, they run quickly and
lightly, and "rise with a sharp click, repeated several times in
quick succession, and move with an easy, undulating motion for
a short distance, when they alight very suddenly, seeming to fall
perpendicularly several feet to the ground." Another peculiarity
of their flight is their habit of flying about in circles, to and fro,
keeping up a constant chirping or call. It is only in the mating
season, when we rarely hear them, that the longspurs have the
angelic manner of singing as they fly, like the skylark. The
colors of the males, among the several longspurs, may differ
widely, but the indistinctly marked females are so like each other
that only their mates, perhaps, could tell them apart.

Lapland Longspur

(Calcarius lapponicus) Finch family

Called also : LAPLAND SNOWBIRD; LAPLAND LARK
BUNTING

Length—6.5 to 7 inches.   A trifle larger than the English sparrow.
Male—Color varies with season. *Winter plumage :* Top of head
    black, with rusty markings, all feathers being tipped with
    white. Behind and below the eye rusty black. Breast and
    underneath grayish white, faintly streaked with black. Above,

SMITH'S LONGSPUR.

reddish brown with black markings. Feet, which are black, have conspicuous, long hind claws or spur.

Female—Rusty gray above, less conspicuously marked. Whitish below.

Range—Circumpolar regions; northern United States; occasional in Middle States; abundant in winter as far as Kansas and the Rocky Mountains.

Migrations—Winter visitors, rarely resident, and without a fixed season.

This arctic bird, although considered somewhat rare with us, when seen at all in midwinter is in such large flocks that, before its visit in the neighborhood is ended, and because there are so few other birds about, it becomes delightfully familiar as it nimbly runs over the frozen ground, picking up grain that has blown about from the barn, when the seeds of the field are buried under snow. This lack of fear through sharp hunger, that often drives the shyest of the birds to our very doors in winter, is as pathetic as it is charming. Possibly it is not so rare a bird as we think, for it is often mistaken for some of the sparrows, the shore larks, and the snow buntings, that it not only resembles, but whose company it frequently keeps, or for one of the other long-spurs.

At all seasons of the year a ground bird, you may readily identify the Lapland longspur by its tracks through the snow, showing the mark of the long hind claw or spur. In summer we know little or nothing about it, for, with the coming of the first flowers, it is off to the far north, where, we are told, it depresses its nest in a bed of moss upon the ground, and lines it with fur shed from the coat of the arctic fox.

Chipping Sparrow

(Spizella socialis) Finch family

Called also : CHIPPY ; HAIR-BIRD ; CHIP-BIRD ; SOCIAL SPARROW

Length—5 to 5.5 inches. An inch shorter than the English sparrow.

Male—Under the eye, on the back of the neck, underneath, and on the lower back ash-gray. Gray stripe over the eye, and a

149

blackish brown one apparently through it. Dark red-brown crown. Back brown, slightly rufous, and feathers streaked with black. Wings and tail dusty brown. Wing-bars not conspicuous. Bill black.

Female—Lacks the chestnut color on the crown, which is streaked with black. In winter the frontlet is black. Bill brownish.

Range—North America, from Newfoundland to the Gulf of Mexico and westward to the Rockies. Winters in Gulf States and Mexico. Most common in eastern United States.

Migrations—April. October. Common summer resident, many birds remaining all the year from southern New England southward.

Who does not know this humblest, most unassuming little neighbor that comes hopping to our very doors ; this mite of a bird with "one talent" that it so persistently uses all the day and every day throughout the summer? Its high, wiry trill, like the buzzing of the locust, heard in the dawn before the sky grows even gray, or in the middle of the night, starts the morning chorus; and after all other voices are hushed in the evening, its tremolo is the last bed-song to come from the trees. But however monotonous such cheerfulness sometimes becomes when we are surfeited with real songs from dozens of other throats, there are long periods of midsummer silence that it punctuates most acceptably.

Its call-note, *chip! chip!* from which several of its popular names are derived, is altogether different from the trill which must do duty as a song to express love, contentment, everything that so amiable a little nature might feel impelled to voice.

But with all its virtues, the chippy shows lamentable weakness of character in allowing its grown children to impose upon it, as it certainly does. In every group of these birds throughout the summer we can see young ones (which we may know by the black line-stripes on their breasts) hopping around after their parents, that are often no larger or more able-bodied than they, and teasing to be fed; drooping their wings to excite pity for a helplessness that they do not possess when the weary little mother hops away from them, and still persistently chirping for food until she weakly relents, returns to them, picks a seed from the ground and thrusts it down the bill of the sauciest teaser in the group. With two such broods in a season the chestnut feathers on the father's jaunty head might well turn gray.

Unlike most of the sparrows, the little chippy frequents high trees, where its nest is built quite as often as in the low bushes of the garden. The horse-hair, which always lines the grassy cup that holds its greenish-blue, speckled eggs, is alone responsible for the name hair-bird, and not the chippy's hair-like trill, as some suppose.

English Sparrow

(Passer domesticus) Finch family

Called also : HOUSE SPARROW

Length—6.33 inches.

Male—Ashy above, with black and chestnut stripes on back and shoulders. Wings have chestnut and white bar, bordered by faint black line. Gray crown, bordered from the eye backward and on the nape by chestnut. Middle of throat and breast black. Underneath grayish white.

Female—Paler; wing-bars indistinct, and without the black marking on throat and breast.

Range—Around the world. Introduced and naturalized in America, Australia, New Zealand.

Migrations—Constant resident.

"Of course, no self-respecting ornithologist will condescend to enlarge his list by counting in the English sparrow—too pestiferous to mention," writes Mr. H. E. Parkhurst, and yet of all bird neighbors is any one more within the scope of this book than the audacious little gamin that delights in the companionship of humans even in their most noisy city thoroughfares ?

In a bulletin issued by the Department of Agriculture it is shown that the progeny of a single pair of these sparrows might amount to 275,716,983,698 in ten years ! Inasmuch as many pairs were liberated in the streets of Brooklyn, New York, in 1851, when the first importation was made, the day is evidently not far off when these birds, by no means meek, "shall inherit the earth."

In Australia Scotch thistles, English sparrows, and rabbits, three most unfortunate importations, have multiplied with equal rapidity until serious alarm fills the minds of the colonists. But in England a special committee appointed by the House of Commons to investigate the character of the alleged pest has yet to learn whether the sparrow's services as an insect-destroyer do not outweigh the injury it does to fruit and grain.

Field Sparrow

(Spizella pusilla) Finch family

Called also: FIELD BUNTING ; WOOD SPARROW ; BUSH SPARROW

Length—5.5 to 5.75 inches. A little smaller than the English sparrow.

Male—Chestnut crown. Upper back bright chestnut, finely streaked with black and ashy brown. Lower back more grayish. Whitish wing-bars. Cheeks, line over the eye, throat, pale brownish drab. Tail long. Underneath grayish white, tinged with palest buff on breast and sides. Bill reddish.

Female—Paler; the crown edged with grayish.

Range—North America, from British provinces to the Gulf, and westward to the plains. Winters from Illinois and Virginia southward.

Migrations—April. November. Common summer resident.

Simply because both birds have chestnut crowns, the field sparrow is often mistaken for the dapper, sociable chippy; and, no doubt because it loves such heathery, grassy pastures as are dear to the vesper sparrow, and has bay wings and a sweet song, these two cousins also are often confused. The field sparrow has a more reddish-brown upper back than any of its small relatives; the absence of streaks on its breast and of the white tail quills so conspicuous in the vesper sparrow's flight, sufficiently differentiate the two birds, while the red bill of the field sparrow is a positive mark of identification.

This bird of humble nature, that makes the scrubby pastures and uplands tuneful from early morning until after sunset, flies away with exasperating shyness as you approach. Alighting on a convenient branch, he lures you on with his clear, sweet song. Follow him, and he only hops about from bush to bush, farther and farther away, singing as he goes a variety of strains, which is one of the bird's peculiarities. The song not only varies in individuals, but in different localities, which may be one reason why no two ornithologists record it alike. Doubtless the chief reason for the amusing differences in the syllables into which the songs of birds are often translated in the books, is that the same notes actually sound differently to different individuals. Thus, to people in Massachusetts the white-throated sparrow seems to say, *"Pea-bod-y, Pea-bod-y, Pea-bod-y!"* while good British subjects beyond the New England border hear him sing quite distinctly, *"Sweet Can-a-da, Can-a-da, Can-a-da!"* But however the opinions as to the syllables of the field sparrow's song may differ, all are agreed as to its exquisite quality, that resembles the vesper sparrow's tender, sweet melody. The song begins with three soft, wild whistles, and ends with a series of trills and quavers that gradually melt away into silence: a serene and restful strain as soothing as a hymn. Like the vesper sparrows, these birds sometimes build a plain, grassy nest, unprotected by overhanging bush, flat upon the ground. Possibly from a prudent fear of field-mice and snakes, the little mother most frequently lays her bluish-white, rufous-marked eggs in a nest placed in a bush of a bushy field. Hence John Burroughs has called the bird the "bush sparrow."

Fox Sparrow

(Passerella ilica) Finch family

Called also: FOX-COLORED SPARROW ; FERRUGINOUS FINCH ; FOXY FINCH

Length—6.5 to 7.25 inches. Nearly an inch longer than the English sparrow.

Male and Female—Upper parts reddish brown, varied with ash-gray, brightest on lower back, wings, and tail. Bluish slate about the head. Underneath whitish; the throat, breast, and sides heavily marked with arrow-heads and oblong dashes of reddish brown and blackish.

Range—Alaska and Manitoba to southern United States. Winters chiefly south of Illinois and Virginia. Occasional stragglers remain north most of the winter.

Migrations—March. November. Most common in the migrations.

There will be little difficulty in naming this largest, most plump and reddish of all the sparrows, whose fox-colored feathers, rather than any malicious cunning of its disposition, are responsible for the name it bears. The male bird is incomparably the finest singer of its gifted family. His faint *tseep* call-note gives no indication of his vocal powers that some bleak morning in early March suddenly send a thrill of pleasure through you. It is the most welcome "glad surprise" of all the spring. Without a preliminary twitter or throat-clearing of any sort, the full, rich, luscious tones, with just a tinge of plaintiveness in them, are poured forth with spontaneous abandon. Such a song at such a time is enough to summon anybody with a musical ear out of doors under the leaden skies to where the delicious notes issue from the leafless shrubbery by the roadside. Watch the singer until the song ends, when he will quite likely descend among the dead leaves on the ground and scratch among them like any barn-yard fowl, but somehow contriving to use both feet at once in the operation, as no chicken ever could. He seems to take special delight in damp thickets, where the insects with which he varies his seed diet are plentiful.

Usually the fox sparrows keep in small, loose flocks, apart by themselves, for they are not truly gregarious ; but they may sometimes be seen travelling in company with their white-throated cousins. They are among the last birds to leave us in the late autumn or winter. Mr. Bicknell says that they seem indisposed to sing unless present in numbers. Indeed, they are little inclined to absolute solitude at any time, for even in the nesting season quite a colony of grassy nurseries may be found in the same meadow, and small companies haunt the roadside shrubbery during the migrations.

Grasshopper Sparrow

(Ammodramus savannarum passerinus) Finch family

Called also : YELLOW-WINGED SPARROW

Length—5 to 5.4 inches. About an inch smaller than the English sparrow.

Male and Female—A cream-yellow line over the eye; centre of crown, shoulders, and lesser wing coverts yellowish. Head blackish; rust-colored feathers, with small black spots on back of the neck; an orange mark before the eye. All other upper parts varied red, brown, cream, and black, with a drab wash. Underneath brownish drab on breast, shading to soiled white, and without streaks. Dusky, even, pointed tail feathers have grayish-white outer margins.

Range—Eastern North America, from British provinces to Cuba. Winters south of the Carolinas.

Migrations—April. October. Common summer resident.

It is safe to say that no other common bird is so frequently overlooked as this little sparrow, that keeps persistently to the grass and low bushes, and only faintly lifts up a weak, wiry voice that is usually attributed to some insect. At the bend of the wings only are the feathers really yellow, and even this bright shade often goes unnoticed as the bird runs shyly through an old dairy field or grassy pasture. You may all but step upon it before it takes wing and exhibits itself on the fence-rail, which is usually as far from the ground as it cares to go. If you are near enough to this perch you may overhear the *zee-e-e-e-e-e-e* that has earned it the name of grasshopper sparrow. If you persistently follow it too closely, away it flies, then suddenly drops to the ground where a scrubby bush affords protection. A curious fact about this bird is that after you have once become acquainted with it, you find that instead of being a rare discovery, as you had supposed, it is apt to be a common resident of almost every field you walk through.

Savanna Sparrow

(*Ammodramus sandwichensis savanna*) Finch family

Called also : SAVANNA BUNTING

Length—5.5 to 6 inches. A trifle smaller than the English sparrow.

Male and Female—Cheeks, space over the eye, and on the bend of the wings pale yellow. General effect of the upper parts brownish drab, streaked with black. Wings and tail dusky, the outer webs of the feathers margined with buff. Under parts white, heavily streaked with blackish and rufous, the marks on breast feathers being wedge-shaped. In the autumn the plumage is often suffused with a yellow tinge.

Range—Eastern North America, from Hudson Bay to Mexico. Winters south of Illinois and Virginia.

Migrations—April. October. A few remain in sheltered marshes at the north all winter.

Look for the savanna sparrow in salt marshes, marshy or upland pastures, never far inland, and if you see a sparrowy bird, unusually white and heavily streaked beneath, and with pale yellow markings about the eye and on the bend of the wing, you may still make several guesses at its identity before the weak, little insect-like trill finally establishes it. Whoever can correctly name every sparrow and warbler on sight is a person to be envied, if, indeed, he exists at all.

In the lowlands of Nova Scotia and, in fact, of all the maritime provinces, this sparrow is the one that is perhaps most commonly seen. Every fence-rail has one perched upon it, singing "*Ptsip, ptsip, ptsip, zee-e-e-e-e*" close to the ear of the passer-by, who otherwise might not hear the low grasshopper-like song. At the north the bird somehow loses the shyness that makes it comparatively little known farther south. Depending upon the scrub and grass to conceal it, you may almost tread upon it before it startles you by its sudden rising with a whirring noise, only to drop to the ground again just as suddenly a few yards farther away, where it scuds among the underbrush and is lost to sight. Tall weeds and fence-rails are as high and exposed situations as it is likely to select while singing. It is most distinctively a ground bird, and flat upon the pasture or in a slightly hollowed cup it has the merest apology for a nest. Only a few wisps of grass are laid in the cavity to receive the pale-green eggs, that are covered most curiously with blotches of brown of many shapes and tints.

Seaside Sparrow

(Ammodramus maritimus) Finch family

Called also : MEADOW CHIPPY; SEASIDE FINCH

Length—6 inches. A shade smaller than the English sparrow.

Male and Female—Upper parts dusky grayish or olivaceous brown, inclining to gray on shoulders and on edges of some feathers. Wings and tail darkest. Throat yellowish white, shading to gray on breast, which is indistinctly mottled and streaked. A yellow spot before the eye and on bend of the wing, the bird's characteristic marks. Blunt tail.

Range—Atlantic seaboard, from Georgia northward. Usually winters south of Virginia.

Migrations—April. November. A few remain in sheltered marshes all winter.

The savanna, the swamp, the sharp-tailed, and the song sparrows may all sometimes be found in the haunts of the seaside sparrow, but you may be certain of finding the latter nowhere else than in the salt marshes within sight or sound of the sea. It is a dingy little bird, with the least definite coloring of all the sparrows that have maritime inclinations, with no rufous tint in its feathers, and less distinct streakings on the breast than any of them. It has no black markings on the back.

Good-sized flocks of seaside sparrows live together in the marshes; but they spend so much of their time on the ground, running about among the reeds and grasses, whose seeds and insect parasites they feed upon, that not until some unusual disturbance in the quiet place flushes them does the intruder suspect their presence. Hunters after beach-birds, longshoremen, seaside cottagers, and whoever follows the windings of a creek through the salt meadows to catch crabs and eels in midsummer, are well acquainted with the "meadow chippies," as the fishermen call them. They keep up a good deal of chirping, sparrow-fashion, and have four or five notes resembling a song that is usually delivered from a tall reed stalk, where the bird sways and balances until his husky performance has ended, when down he drops upon the ground out of sight. Sometimes, too, these notes are uttered while the bird flutters in the air above the tops of the sedges.

Sharp-tailed Sparrow

(*Ammodramus caudacutus*) Finch family

Length—5.25 to 5.85 inches. A trifle smaller than the English sparrow.

Male and Female—Upper parts brownish or grayish olive, the back with black streaks, and gray edges to some feathers. A gray line through centre of crown, which has maroon stripes; gray ears enclosed by buff lines, one of which passes through the eye and one on side of throat; brownish orange, or buff, on sides of head. Bend of the wing yellow. Breast and sides pale buff, distinctly streaked with black. Underneath whitish. Each narrow quill of tail is sharply pointed, the outer ones shortest.

Range—Atlantic coast. Winters south of Virginia.

Migrations—April. November. Summer resident.

This bird delights in the company of the dull-colored seaside sparrow, whose haunts in the salt marshes it frequents, especially the drier parts; but its pointed tail-quills and more distinct markings are sufficient to prevent confusion. Mr. J. Dwight, Jr., who has made a special study of maritime birds, says of it: "It runs about among the reeds and grasses with the celerity of a mouse, and it is not apt to take wing unless closely pressed." (Wilson credited it with the nimbleness of a sandpiper.) "It builds its nest in the tussocks on the bank of a ditch, or in the drift left by the tide, rather than in the grassier sites chosen by its neighbors, the seaside sparrows."

Only rarely does one get a glimpse of this shy little bird, that darts out of sight like a flash at the first approach. Balancing on a cat-tail stalk or perched upon a bit of driftwood, it makes a feeble, husky attempt to sing a few notes; and during the brief performance the opera-glasses may search it out successfully. While it feeds upon the bits of sea-food washed ashore to the edge of the marshes, it gives us perhaps the best chance we ever get, outside of a museum, to study the bird's characteristics of plumage.

"Both the sharp-tailed and the seaside finches are crepuscular," says Dr. Abbott, in "The Birds About Us." They run up and down the reeds and on the water's edge long after most birds have gone to sleep.

Song Sparrow

(Melospiza fasciata) Finch family

Length—6 to 6.5 inches. About the same size as the English sparrow.

Male and Female—Brown head, with three longitudinal gray bands. Brown stripe on sides of throat. Brownish-gray back, streaked with rufous. Underneath gray, shading to white, heavily streaked with darkest brown. A black spot on breast. Wings without bars. Tail plain grayish brown.

Range—North America, from Fur Countries to the Gulf States. Winters from southern Illinois and Massachusetts to the Gulf.

Migrations—March. November. A few birds remain at the north all the year.

Here is a veritable bird neighbor, if ever there was one ; at home in our gardens and hedges, not often farther away than the

SONG SPARROW.
⅓ Life-size.

roadside, abundant everywhere during nearly every month in the year, and yet was there ever one too many ? There is scarcely an hour in the day, too, when its delicious, ecstatic song may not be heard ; in the darkness of midnight, just before dawn, when its voice is almost the first to respond to the chipping sparrow's wiry trill and the robin's warble ; in the cool of the morning, the heat of noon, the hush of evening—ever the simple, homely, sweet melody that every good American has learned to love in childhood. What the bird lacks in beauty it abundantly makes up in good cheer. Not at all retiring, though never bold, it chooses some conspicuous perch on a bush or tree to deliver its outburst of song, and sings away with serene unconsciousness. Its artlessness is charming. Thoreau writes in his "Summer" that the country girls in Massachusetts hear the bird say : " *Maids, maids, maids, hang on your teakettle, teakettle-ettle-ettle.*" The call-note, a metallic *chip*, is equally characteristic of the bird's irrepressible vivacity. It has still another musical expression, however, a song more prolonged and varied than its usual performance, that it seems to sing only on the wing.

Of course, the song sparrow must sometimes fly upward, but whoever sees it fly anywhere but downward into the thicket that it depends upon to conceal it from too close inspection ? By pumping its tail as it flies, it seems to acquire more than the ordinary sparrow's velocity.

Its nest, which is likely to be laid flat on the ground, except where field-mice are plentiful (in which case it is elevated into the crotch of a bush), is made of grass, strips of bark, and leaves, and lined with finer grasses and hair. Sometimes three broods may be reared in a season, but even the cares of providing insects and seeds enough for so many hungry babies cannot altogether suppress the cheerful singer. The eggs are grayish white, speckled and clouded with lavender and various shades of brown.

In sparsely settled regions the song sparrows seem to show a fondness for moist woodland thickets, possibly because their tastes are insectivorous. But it is difficult to imagine the friendly little musician anything but a neighbor.

Swamp Song Sparrow

(Melospiza georgiana) Finch family

Called also : SWAMP SPARROW; MARSH SPARROW; RED GRASS-BIRD; SWAMP FINCH

Length—5 to 5.8 inches. A little smaller than the English sparrow.

Male—Forehead black; crown, which in winter has black stripes, is always bright bay; line over the eye, sides of the neck gray. Back brown, striped with various shades. Wing-edges and tail reddish brown. Mottled gray underneath, inclining to white on the chin.

Female—Without black forehead and stripes on head.

Range—North America, from Texas to Labrador.

Migrations—April. October. A few winter at the north.

In just such impenetrable retreats as the marsh wrens choose, another wee brown bird may sometimes be seen springing up from among the sedges, singing a few sweet notes as it flies and floats above them, and then suddenly disappearing into the grassy tangle. It is too small, and its breast is not streaked enough to be a song sparrow, neither are their songs alike; it has not the wren's peculiarities of bill and tail. Its bright-bay crown and sparrowy markings finally identify it. A suggestion of the bird's watery home shows itself in the liquid quality of its simple, sweet note, stronger and sweeter than the chippy's, and repeated many times almost like a trill that seems to trickle from the marsh in a little rivulet of song. The sweetness is apt to become monotonous to all but the bird itself, that takes evident delight in its performance. In the spring, when flocks of swamp sparrows come north, how they enliven the marshes and waste places! And yet the song, simple as it is, is evidently not uttered altogether without effort, if the tail-spreading and teetering of the body after the manner of the ovenbird, are any indications of exertion.

Nuttall says of these birds: "They thread their devious way with the same alacrity as the rail, with whom, indeed, they are often associated in neighborhood. In consequence of this perpetual brushing through sedge and bushes, their feathers are frequently so worn that their tails appear almost like those of rats."

But the swamp sparrows frequently belie their name, and, especially in the South, live in dry fields, worn-out pasture lands with scrubby, weedy patches in them. They live upon seeds of grasses and berries, but Dr. Abbott has detected their special fondness for fish—not fresh fish particularly, but rather such as have lain in the sun for a few days and become dry as a chip.

Their nest is placed on the ground, sometimes in a tussock of grass or roots of an upturned tree quite surrounded by water. Four or five soiled white eggs with reddish-brown spots are laid usually twice in a season.

Tree Sparrow

(Spizella monticola) Finch family

Called also : CANADA SPARROW ; WINTER CHIPPY ; TREE BUNTING ; WINTER CHIP-BIRD ; ARCTIC CHIPPER

Length—6 to 6.35 inches. About the same size as the English sparrow.

Male—Crown of head bright chestnut. Line over the eye, cheeks, throat, and breast gray, the breast with an indistinct black spot on centre. Brown back, the feathers edged with black and buff. Lower back pale grayish brown. Two whitish bars across dusky wings; tail feathers bordered with grayish white. Underneath whitish.

Female—Smaller and less distinctly marked.

Range—North America, from Hudson Bay to the Carolinas, and westward to the plains.

Migrations—October. April. Winter resident.

A revised and enlarged edition of the friendly little chipping sparrow, that hops to our very doors for crumbs throughout the mild weather, comes out of British America at the beginning of winter to dissipate much of the winter's dreariness by his cheerful twitterings. Why he should have been called a tree sparrow is a mystery, unless because he does not frequent trees—a reason with sufficient plausibility to commend the name to several of the early ornithologists, who not infrequently called a bird precisely what it was not. The tree sparrow actually does not show half the preference for trees that its familiar little counterpart does, but rather keeps to low bushes when not on the ground, where we usually find it. It does not crouch upon the ground like the chippy, but with a lordly carriage holds itself erect as it nimbly runs over the frozen crust. Sheltered from the high, wintry winds in the furrows and dry ditches of ploughed fields, a loose flock of these active birds keep up a merry hunt for fallen seeds and berries, with a belated beetle to give the grain a relish. As you approach the feeding ground, one bird gives a shrill alarm-cry, and instantly five times as many birds as you suspected were in the field take wing and settle down in the scrubby undergrowth at the edge of the woods or by the wayside. No still cold seems too keen for them to go a-foraging; but when cutting winds blow through the leafless thickets the scattered remnants of a flock seek the shelter of stone walls, hedges, barns, and cozy nooks about the house and garden. It is in midwinter that these birds grow most neighborly, although even then they are distinctly less sociable than their small chippy cousins.

By the first of March, when the fox sparrow and the bluebird attract the lion's share of attention by their superior voices, we not infrequently are deaf to the modest, sweet little strain that answers for the tree sparrow's love-song. Soon after the bird is in full voice, away it goes with its flock to their nesting ground in Labrador or the Hudson Bay region. It builds, either on the ground or not far from it, a nest of grasses, rootlets, and hair, without which no true chippy counts its home complete.

Vesper Sparrow

(Poocætes gramineus) Finch family

Called also: BAY-WINGED BUNTING; GRASSFINCH; GRASS-
BIRD

Length—5.75 to 6.25 inches. A little smaller than the English sparrow.
Male and Female—Brown above, streaked and varied with gray. Lesser wing coverts bright rufous. Throat and breast whitish, striped with dark brown. Underneath plain soiled white. Outer tail-quills, which are its special mark of identification, are partly white, but apparently wholly white as the bird flies.

Range—North America, especially common in eastern parts from Hudson Bay to Gulf of Mexico. Winters south of Virginia.
Migrations—April. October. Common summer resident.

Among the least conspicuous birds, sparrows are the easiest to classify for that very reason, and certain prominent features of the half dozen commonest of the tribe make their identification simple even to the merest novice. The distinguishing marks of this sparrow that haunts open, breezy pasture lands and country waysides are its bright, reddish-brown wing coverts, prominent among its dingy, pale brownish-gray feathers, and its white tail-quills, shown as the bird flies along the road ahead of you to light upon the fence-rail. It rarely flies higher, even to sing its serene, pastoral strain, restful as the twilight, of which, indeed, it seems to be the vocal expression. How different from the ecstatic outburst of the song sparrow ! Pensive, but not sad, its long-drawn silvery notes continue in quavers that float off unended like a trail of mist. The song is suggestive of the thoughts that must come at evening to some New England saint of humble station after a well-spent, soul-uplifting day.

But while the vesper sparrow sings oftenest and most sweetly in the late afternoon and continues singing until only he and the rose-breasted grosbeak break the silence of the early night, his is one of the first voices to join the morning chorus. No "early worm," however, tempts him from his grassy nest, for the seeds in the pasture lands and certain tiny insects that live among the grass furnish meals at all hours. He simply delights in the cool, still morning and evening hours and in giving voice to his enjoyment of them.

The vesper sparrow is preëminently a grass-bird. It first opens its eyes on the world in a nest neatly woven of grasses, laid on the ground among the grass that shelters it and furnishes it with food and its protective coloring. Only the grazing cattle know how many nests and birds are hidden in their pastures. Like the meadowlarks, their presence is not even suspected until a flock is flushed from its feeding ground, only to return to the spot when you have passed on your way. Like the meadowlark again, the vesper sparrow occasionally sings as it soars upward from its grassy home.

White-crowned Sparrow

(Zonotrichia leucophrys) Finch family

Length—7 inches. A little larger than the English sparrow.

Male—White head, with four longitudinal black lines marking off a crown, the black-and-white stripes being of about equal width. Cheeks, nape, and throat gray. Light gray underneath, with some buff tints. Back dark grayish brown, some feathers margined with gray. Two interrupted white bars across wings. Plain, dusky tail ; total effect, a clear ashen gray.

Female—With rusty head inclining to gray on crown. Paler throughout than the male.

Range—From high mountain ranges of western United States (more rarely on Pacific slope) to Atlantic Ocean, and from Labrador to Mexico. Chiefly south of Pennsylvania.

Migrations—October. April. Irregular migrant in Northern States. A winter resident elsewhere.

The large size and handsome markings of this aristocratic-looking Northern sparrow would serve to distinguish him at once, did he not often consort with his equally fine-looking white-throated cousins while migrating, and so too often get overlooked. Sparrows are such gregarious birds that it is well to scrutinize every flock with especial care in the spring and autumn, when the rarer migrants are passing. This bird is more common in the high altitudes of the Sierra Nevada and Rocky Mountains than elsewhere in the United States. There in the lonely forest it nests in low bushes or on the ground, and sings its full love-song, as it does in the northern British provinces, along the Atlantic coast ; but during the migrations it favors us only with selections from its repertoire. Mr. Ernest Thompson says, "Its usual song is like the latter half of the white-throat's familiar refrain, repeated a number of times with a peculiar, sad cadence and in a clear, soft whistle that is characteristic of the group." "The song is the loudest and most plaintive of all the sparrow songs," says John Burroughs. "It begins with the words *fe-u, fe-u, fe-u,* and runs off into trills and quavers like the song sparrow's, only much more touching." Colorado miners tell that this sparrow, like its white-throated relative, sings on the darkest nights. Often a score or more birds are heard singing at once after the habit of the European nightingales, which, however, choose to sing only in the moonlight.

White-throated Sparrow

(Zonotrichia albicollis) Finch family

Called also: PEABODY BIRD; CANADA SPARROW

Length—6.75 to 7 inches. Larger than the English sparrow.
Male and Female—A black crown divided by narrow white line. Yellow spot before the eye, and a white line, apparently running through it, passes backward to the nape. Conspicuous white throat. Chestnut back, varied with black and whitish. Breast gray, growing lighter underneath. Wings edged with rufous and with two white cross-bars.
Range—Eastern North America. Nests from Michigan and Massachusetts northward to Labrador. Winters from southern New England to Florida.
Migrations—April. October. Abundant during migrations, and in many States a winter resident.

"*I-I, Pea-body, Pea-body, Pea-body,*" are the syllables of the white-throat's song heard by the good New Englanders, who have a tradition that you must either be a Peabody or a nobody there; while just over the British border the bird is distinctly understood to say, "*Swee-e-e-t Can-a-da, Can-a-da, Can-a-da.*" "*All day, whit-tle-ing, whit-tle-ing, whit-tle-ing,*" the Maine people declare he sings; and Hamilton Gibson told of a perplexed farmer, Peverly by name, who, as he stood in the field undecided as to what crop to plant, clearly heard the bird advise, "*Sow wheat, Pev-er-ly, Pev-er-ly, Pev-er-ly.*" Such divergence of opinion, which is really slight compared with the verbal record of many birds' songs, only goes to show how little the sweetness of birds' music, like the perfume of a rose, depends upon a name.

In a family not distinguished for good looks, the white-throated sparrow is conspicuously handsome, especially after the spring moult. In midwinter the feathers grow dingy and the markings indistinct ; but as the season advances, his colors are sure to brighten perceptibly, and before he takes the northward journey in April, any little lady sparrow might feel proud of the attentions of so fine-looking and sweet-voiced a lover. The black, white, and yellow markings on his head are now clear and beautiful.   His figure is plump and aristocratic.

These sparrows are particularly sociable travellers, and cordially welcome many stragglers to their flocks—not during the migrations only, but even when winter's snow affords only the barest gleanings above it.   Then they boldly peck about the dog's plate by the kitchen door and enter the barn-yard, calling their feathered friends with a sharp *tseep* to follow them.   Seeds and insects are their chosen food, and were they not well wrapped in an adipose coat under their feathers, there must be many a winter night when they would go shivering, supperless, to their perch.

In the dark of midnight one may sometimes hear the white-throat softly singing in its dreams.

GREEN, GREENISH GRAY, OLIVE, AND YELLOWISH OLIVE BIRDS

Tree Swallow
Ruby-throated Humming-bird
Golden-crowned Kinglet
Ruby-crowned Kinglet
Solitary Vireo
Red-eyed Vireo
White-eyed Vireo
Warbling Vireo
Ovenbird
Worm-eating Warbler
Acadian Flycatcher
Yellow-bellied Flycatcher
Black-throated Green Warbler

Look also among the Olive-brown Birds, especially for the Cuckoos, Alice's and the Olive-backed Thrushes; and look in the yellow group, many of whose birds are olive also. See also females of the Red Crossbill, Orchard Oriole, Scarlet Tanager, Summer Tanager.

GREEN, GREENISH GRAY, OLIVE, AND YELLOWISH OLIVE BIRDS

Tree Swallow

(Tachycineta bicolor) Swallow family

Called also : WHITE-BELLIED SWALLOW

Length—5 to 6 inches. A little shorter than the English sparrow, but apparently much larger because of its wide wing-spread.

Male—Lustrous dark steel-green above; darker and shading into black on wings and tail, which is forked. Under parts soft white.

Female—Duller than male.

Range—North America, from Hudson Bay to Panama.

Migrations—End of March. September or later. Summer resident.

"The stork in the heaven knoweth her appointed times; and the turtle and the crane and the swallow observe the time of their coming."—Jeremiah, viii. 7.

The earliest of the family to appear in the spring, the tree swallow comes skimming over the freshly ploughed fields with a wide sweep of the wings, in what appears to be a perfect ecstasy of flight. More shy of the haunts of man, and less gregarious than its cousins, it is usually to be seen during migration flying low over the marshes, ponds, and streams with a few chosen friends, keeping up an incessant warbling twitter while performing their bewildering and tireless evolutions as they catch their food on the wing. Their white breasts flash in the sunlight, and it is only when they dart near you, and skim close along the surface of the water, that you discover their backs to be not black, but rich, dark green, glossy to iridescence.

It is probable that these birds keep near the waterways because their favorite insects and wax-berries are more plentiful in such places; but this peculiarity has led many people to the absurd belief that the tree swallow buries itself under the mud of ponds in winter in a state of hibernation. No bird's breathing apparatus is made to operate under mud.

In unsettled districts these swallows nest in hollow trees, hence their name; but with that laziness that forms a part of the degeneracy of civilization, they now gladly accept the boxes about men's homes set up for the martins. Thousands of these beautiful birds have been shot on the Long Island marshes and sold to New York epicures for snipe.

Ruby-throated Humming-bird

(Trochilus colubris) Humming-bird family

Length—3.5 to 3.75 inches. A trifle over half as long as the English sparrow. The smallest bird we have.

Male—Bright metallic green above; wings and tail darkest, with ruddy-purplish reflections and dusky-white tips on outer tail-quills. Throat and breast brilliant metallic-red in one light, orange flame in another, and dusky orange in another, according as the light strikes the plumage. Sides greenish; underneath lightest gray, with whitish border outlining the brilliant breast. Bill long and needle-like.

Female—Without the brilliant feathers on throat; darker gray beneath. Outer tail-quills are banded with black and tipped with white.

Range—Eastern North America, from northern Canada to the Gulf of Mexico in summer. Winters in Central America.

Migrations—May. October. Common summer resident.

This smallest, most exquisite and unabashed of our bird neighbors cannot be mistaken, for it is the only one of its kin found east of the plains and north of Florida, although about four hundred species, native only to the New World, have been named by scientists. How does it happen that this little tropical jewel alone flashes about our Northern gardens? Does it never stir the spirit of adventure and emulation in the glistening breasts of its stay-at-home cousins in the tropics by tales of luxuriant tangles of honeysuckle and clematis on our cottage porches; of deep-cupped trumpet-flowers climbing over the walls of old-fashioned gardens, where larkspur, narcissus, roses, and phlox, that crowd the box-edged beds, are more gay and honey-laden than their

RUBY-THROATED HUMMING BIRDS.
Life-size.

little brains can picture? Apparently it takes only the wish to be in a place to transport one of these little fairies either from the honeysuckle trellis to the canna bed or from Yucatan to the Hudson. It is easy to see how to will and to fly are allied in the minds of the humming-birds, as they are in the Latin tongue. One minute poised in midair, apparently motionless before a flower while draining the nectar from its deep cup—though the humming of its wings tells that it is suspended there by no magic —the next instant it has flashed out of sight as if a fairy's wand had made it suddenly invisible. Without seeing the hummer, it might be, and often is, mistaken for a bee improving the "shining hour."

At evening one often hears of a "humming-bird" going the rounds of the garden, but at this hour it is usually the sphinx-moth hovering above the flower-beds—the one other creature besides the bee for which the bird is ever mistaken. The postures and preferences of this beautiful large moth make the mistake a very natural one.

The ruby-throat is strangely fearless and unabashed. It will dart among the vines on the veranda while the entire household are assembled there, and add its hum to that of the conversation in a most delightfully neighborly way. Once a glistening little sprite, quite undaunted by the size of an audience that sat almost breathless enjoying his beauty, thrust his bill into one calyx after another on a long sprig of honeysuckle held in the hand.

And yet, with all its friendliness—or is it simply fearlessness? —the bird is a desperate duellist, and will longe his deadly blade into the jewelled breast of an enemy at the slightest provocation and quicker than thought. All the heat of his glowing throat seems to be transferred to his head while the fight continues, sometimes even to the death—a cruel, but marvellously beautiful sight as the glistening birds dart and tumble about beyond the range of peace-makers.

High up in a tree, preferably one whose knots and lichen-covered excrescences are calculated to help conceal the nest that so cleverly imitates them, the mother humming-bird saddles her exquisite cradle to a horizontal limb. She lines it with plant-down, fluffy bits from cat-tails, and the fronds of fern, felting the material into a circle that an elm-leaf amply roofs over. Outside, lichens or bits of bark blend the nest so harmoniously with its surroundings that one may look long and thoroughly before discovering it. Two infinitesimal, white eggs tax the nest accommodation to its utmost.

In the mating season the female may be seen perching—a posture one rarely catches her gay lover in—preening her dainty but sombre feathers with ladylike nicety. The young birds do a great deal of perching before they gain the marvellously rapid wing-motions of maturity, but they are ready to fly within a week after they are hatched. By the time the trumpet-vine is in bloom they dart and sip and utter a shrill little squeak among the flowers, in company with the old birds.

During the nest-building and incubation the male bird keeps so aggressively on the defensive that he often betrays to a hitherto unsuspecting intruder the location of his home. After the young birds have to be fed he is most diligent in collecting food, that consists not alone of the sweet juices of flowers, as is popularly supposed, but also of aphides and plant-lice that his proboscis-like tongue licks off the garden foliage literally like a streak of lightning.

Both parents feed the young by regurgitation—a process disgusting to the human observer, whose stomach involuntarily revolts at the sight so welcome to the tiny, squeaking, hungry birds.

Ruby-crowned Kinglet

(Regulus calendula) Kinglet family

Called also: RUBY-CROWNED WREN ; RUBY-CROWNED WARBLER

Length—4.25 to 4.5 inches. About two inches smaller than the English sparrow.

Male—Upper parts grayish olive-green, brighter nearer the tail; wings and tail dusky, edged with yellowish olive. Two whitish wing-bars. Breast and underneath light yellowish gray. In the adult male a vermilion spot on crown of his ash-gray head.

Female—Similar, but without the vermilion crest.

Range—North America. Breeds from northern United States northward. Winters from southern limits of its breeding range to Central America and Mexico.

Migrations—October. April. Rarely a winter resident at the North. Most common during its migrations.

RUBY-CROWNED KINGLET.
Life-size.

A trifle larger than the golden-crowned kinglet, with a vermilion crest instead of a yellow and flame one, and with a decided preference for a warmer winter climate, and the ruby-crown's chief distinguishing characteristics are told. These rather confusing relatives would be less puzzling if it were the habit of either to keep quiet long enough to focus the opera-glasses on their crowns, which it only rarely is while some particularly promising haunt of insects that lurk beneath the rough bark of the evergreens has to be thoroughly explored. At all other times both kinglets keep up an incessant fluttering and twinkling among the twigs and leaves at the ends of the branches, jerking their tiny bodies from twig to twig in the shrubbery, hanging head downward, like a nuthatch, and most industriously feeding every second upon the tiny insects and larvæ hidden beneath the bark and leaves. They seem to be the feathered expression of perpetual motion. And how dainty and charming these tiny sprites are! They are not at all shy; you may approach them quite close if you will, for the birds are simply too intent on their business to be concerned with yours.

If a sharp lookout be kept for these ruby-crowned migrants, that too often slip away to the south before we know they have come, we notice that they appear about a fortnight ahead of the golden-crested species, since the mild, soft air of our Indian summer is exactly to their liking. At this season there is nothing in the bird's "thin, metallic call-note, like a vibrating wire," to indicate that he is one of our finest songsters. But listen for him during the spring migration, when a love-song is already ripening in his tiny throat. What a volume of rich, lyrical melody pours from the Norway spruce, where the little musician is simply practising to perfect the richer, fuller song that he sings to his nesting mate in the far north! The volume is really tremendous, coming from so tiny a throat. Those who have heard it in northern Canada describe it as a flute-like and mellow warble full of intricate phrases past the imitating. Dr. Coues says of it: "The kinglet's exquisite vocalization defies description."

Curiously enough, the nest of this bird, that is not at all rare, has been discovered only six times. It would appear to be overlarge for the tiny bird, until we remember that kinglets are wont to have a numerous progeny in their pensile, globular home. It is made of light, flimsy material—moss, strips of bark, and plant-fibre well knit together and cosily lined with feathers, which must be a grateful addition to the babies, where they are reared in evergreens in cold, northern woods.

Golden-crowned Kinglet

(Regulus satrapa) Kinglet family

Called also: GOLDEN-CROWNED GOLDCREST; FIERY-CROWNED WREN

Length—4 to 4.25 inches. About two inches smaller than the English sparrow.

Male—Upper parts grayish olive-green; wings and tail dusky, margined with olive-green. Underneath soiled whitish. Centre of crown bright orange, bordered by yellow and enclosed by black line. Cheeks gray; a whitish line over the eye.

Female—Similar, but centre of crown lemon-yellow and more grayish underneath.

Range—North America generally. Breeds from northern United States northward. Winters chiefly from North Carolina to Central America, but many remain north all the year.

Migrations—September. April. Chiefly a winter resident south of Canada.

If this cheery little winter neighbor would keep quiet long enough, we might have a glimpse of the golden crest that distinguishes him from his equally lively cousin, the ruby-crowned; but he is so constantly flitting about the ends of the twigs, peering at the bark for hidden insects, twinkling his wings and fluttering among the evergreens with more nervous restlessness than a vireo, that you may know him well before you have a glimpse of his tri-colored crown.

When the autumn foliage is all aglow with yellow and flame this tiny sprite comes out of the north, where neither nesting nor moulting could rob him of his cheerful spirits. Except the humming-bird and the winter wren, he is the smallest bird we have. And yet, somewhere stored up in his diminutive body, is warmth enough to withstand zero weather. With evident enjoyment of the cold, he calls out a shrill, wiry *zee, zee, zee*, that rings merrily from the pines and spruces when our fingers are too numb to hold the opera-glasses in an attempt to follow his restless flittings from branch to branch. Is it one of the unwritten laws of birds that the smaller their bodies the greater their activity ?

When you see one kinglet about, you may be sure there are others not far away, for, except in the nesting season, its habits are distinctly social, its friendliness extending to the humdrum brown creeper, the chickadees, and the nuthatches, in whose company it is often seen ; indeed, it is likely to be in almost any flock of the winter birds. They are a merry band as they go exploring the trees together. The kinglet can hang upside down, too, like the other acrobats, many of whose tricks he has learned ; and it can pick off insects from a tree with as business-like an air as the brown creeper, but with none of that soulless bird's plodding precision.

In the early spring, just before this busy little sprite leaves us to nest in Canada or Labrador—for heat is the one thing that he can't cheerfully endure—a gushing, lyrical song bursts from his tiny throat—a song whose volume is so out of proportion to the bird's size that Nuttall's classification of kinglets with wrens doesn't seem far wrong after all.

Only rarely is a nest found so far south as the White Mountains. It is said to be extraordinarily large for so small a bird ; but that need not surprise us when we learn that as many as ten creamy-white eggs, blotched with brown and lavender, are no uncommon number for the pensile cradle to hold. How do the tiny parents contrive to cover so many eggs and to feed such a nestful of fledglings?

Solitary Vireo

(Vireo solitarius) Vireo or Greenlet family

Called also : BLUE-HEADED VIREO

Length—5.5 to 7 inches. A little smaller than the English sparrow.

Male—Dusky olive above ; head bluish gray, with a white line around the eye, spreading behind the eye into a patch. Beneath whitish, with yellow-green wash on the sides. Wings dusky olive, with two distinct white bars. Tail dusky, some quills edged with white.

Female—Similar, but her head is dusky olive.

Range—United States to plains, and the southern British provinces. Winters in Florida and southward.

Migrations—May. Early October. Common during migrations; more rarely a summer resident south of Massachusetts.

By no means the recluse that its name would imply, the solitary vireo, while a bird of the woods, shows a charming curiosity about the stranger with opera-glasses in hand, who has penetrated to the deep, swampy tangles, where it chooses to live. Peering at you through the green undergrowth with an eye that seems especially conspicuous because of its encircling white rim, it is at least as sociable and cheerful as any member of its family, and Mr. Bradford Torrey credits it with "winning tameness." "Wood-bird as it is," he says, "it will sometimes permit the greatest familiarities. Two birds I have seen, which allowed themselves to be stroked in the freest manner, while sitting on the eggs, and which ate from my hand as readily as any pet canary."

The solitary vireo also builds a pensile nest, swung from the crotch of a branch, not so high from the ground as the yellow-throated vireo's nor so exquisitely finished, but still a beautiful little structure of pine-needles, plant-fibre, dry leaves, and twigs, all lichen-lined and bound and rebound with coarse spiders' webs.

The distinguishing quality of this vireo's celebrated song is its tenderness : a pure, serene uplifting of its loving, trustful nature that seems inspired by a fine spirituality.

Red-eyed Vireo

(*Vireo olivaceus*) Vireo or Greenlet family

Called also : THE PREACHER

Length—5.75 to 6.25 inches. A fraction smaller than the English sparrow.

Male and Female—Upper parts light olive-green; well-defined slaty-gray cap, with black marginal line, below which, and forming an exaggerated eyebrow, is a line of white. A brownish band runs from base of bill through the eye. The iris is ruby-red. Underneath white, shaded with light greenish yellow on sides and on under tail and wing coverts.

Range—United States to Rockies and northward. Winters in Central and South America.

Migrations—April. October. Common summer resident.

176

"You see it—you know it—do you hear me? Do you believe it?" is Wilson Flagg's famous interpretation of the song of this commonest of all the vireos, that you cannot mistake with such a key. He calls the bird the preacher from its declamatory style: an up-and-down warble delivered with a rising inflection at the close and followed by an impressive silence, as if the little green orator were saying, "I pause for a reply."

Notwithstanding its quiet coloring, that so closely resembles the leaves it hunts among, this vireo is rather more noticeable than its relatives because of its slaty cap and the black-and-white lines over its ruby eye, that, in addition to the song, are its marked characteristics.

Whether she is excessively stupid or excessively kind, the mother-vireo has certainly won for herself no end of ridicule by allowing the cowbird to deposit a stray egg in the exquisitely made, pensile nest, where her own tiny white eggs are lying; and though the young cowbird crowd and worry her little fledglings and eat their dinner as fast as she can bring it in, no displeasure or grudging is shown towards the dusky intruder that is sure to upset the rightful heirs out of the nest before they are able to fly.

In the heat of a midsummer noon, when nearly every other bird's voice is hushed, and only the locust seems to rejoice in the fierce sunshine, the little red-eyed vireo goes persistently about its business of gathering insects from the leaves, not flitting nervously about like a warbler, or taking its food on the wing like a flycatcher, but patiently and industriously dining where it can, and singing as it goes.

When a worm is caught it is first shaken against a branch to kill it before it is swallowed. Vireos haunt shrubbery and trees with heavy foliage, all their hunting, singing, resting, and home-building being done among the leaves—never on the ground.

White-eyed Vireo

(Vireo noveboracensis) Vireo or Greenlet family

Length—5 to 5.3 inches. An inch shorter than the English sparrow.

Male and Female—Upper parts bright olive-green, washed with grayish. Throat and underneath white; the breast and sides greenish yellow; wings have two distinct bars of yellowish white. Yellow line from beak to and around the eye, which has a white iris. Feathers of wings and tail brownish and edged with yellow.

Range—United States to the Rockies, and to the Gulf regions and beyond in winter.

Migrations—May. September. Summer resident.

"Pertest of songsters," the white-eyed vireo makes whatever neighborhood it enters lively at once. Taking up a residence in the tangled shrubbery or thickety undergrowth, it immediately begins to scold like a crotchety old wren. It becomes irritated over the merest trifles—a passing bumblebee, a visit from another bird to its tangle, an unsuccessful peck at a gnat—anything seems calculated to rouse its wrath and set every feather on its little body a-trembling, while it sharply snaps out what might perhaps be freely constructed into "cuss-words."

And yet the inscrutable mystery is that this virago meekly permits the lazy cowbird to deposit an egg in its nest, and will patiently sit upon it, though it is as large as three of her own tiny eggs; and when the little interloper comes out from his shell the mother-bird will continue to give it the most devoted care long after it has shoved her poor little starved babies out of the nest to meet an untimely death in the smilax thicket below.

An unusual variety of expression distinguishes this bird's voice from the songs of the other vireos, which are apt to be monotonous, as they are incessant. If you are so fortunate to approach the white-eyed vireo before he suspects your presence, you may hear him amusing himself by jumbling together snatches of the songs of the other birds in a sort of potpourri; or perhaps he will be scolding or arguing with an imaginary foe, then dropping his voice and talking confidentially to himself. Suddenly he bursts into a charming, simple little song, as if the introspection had given him reason for real joy. All these vocal accomplishments suggest the chat at once; but the minute your intrusion is discovered the sharp scolding, that is fairly screamed at you from an enraged little throat, leaves no possible shadow of a doubt as to the bird you have disturbed. It has the most emphatic call and song to be heard in the woods; it snaps its words off very short. "*Chick-a-rer chick*" is its usual call-note, jerked out with great spitefulness.

178

WARBLING VIREO.
Life-size.

Wilson thus describes the jealously guarded nest: "This bird builds a very neat little nest, often in the figure of an inverted cone; it is suspended by the upper end of the two sides, on the circular bend of a prickly vine, a species of smilax, that generally grows in low thickets. Outwardly it is constructed of various light materials, bits of rotten wood, fibres of dry stalks, of weeds, pieces of paper (commonly newspapers, an article almost always found about its nest, so that some of my friends have given it the name of the politician); all these materials are interwoven with the silk of the caterpillars, and the inside is lined with fine, dry grass and hair."

Warbling Vireo

(Vireo gilvus) Vireo or Greenlet family

Length—5.5 to 6 inches. A little smaller than the English sparrow.

Male and Female—Ashy olive-green above, with head and neck ash-colored. Dusky line over the eye. Underneath whitish, faintly washed with dull yellow, deepest on sides ; no bars on wings.

Range—North America, from Hudson Bay to Mexico.

Migrations—May. Late September or early October. Summer resident.

This musical little bird shows a curious preference for rows of trees in the village street or by the roadside, where he can be sure of an audience to listen to his rich, continuous warble. There is a mellowness about his voice, which rises loud, but not altogether cheerfully, above the bird chorus, as if he were a gifted but slightly disgruntled contralto. Too inconspicuously dressed, and usually too high in the tree-top to be identified without opera-glasses, we may easily mistake him by his voice for one of the warbler family, which is very closely allied to the vireos. Indeed, this warbling vireo seems to be the connecting link between them.

Morning and afternoon, but almost never in the evening, we may hear him rippling out song after song as he feeds on insects and berries about the garden. But this familiarity lasts only until nesting time, for off he goes with his little mate to some unfrequented lane near a wood until their family is reared, when, with a perceptibly happier strain in his voice, he once more haunts our garden and row of elms before taking the southern journey.

Ovenbird

(Seiurus aurocapillus) Wood Warbler family

Called also : GOLDEN-CROWNED THRUSH, THE TEACHER; WOOD WAGTAIL ; GOLDEN-CROWNED WAGTAIL ; GOLDEN-CROWNED ACCENTOR

Length—6 to 6.15 inches. Just a shade smaller than the English sparrow.

Male and Female—Upper parts olive, with an orange-brown crown, bordered by black lines that converge toward the bill. Under parts white; breast spotted and streaked on the sides. White eye-ring.

Range—United States, to Pacific slope.

Migrations—May. October. Common summer resident.

Early in May you may have the good fortune to see this little bird of the woods strutting in and out of the garden shrubbery with a certain mock dignity, like a child wearing its father's boots. Few birds can walk without appearing more or less ridiculous, and however gracefully and prettily it steps, this amusing little wagtail is no exception. When seen at all—which is not often, for it is shy—it is usually on the ground, not far from the shrubbery or a woodland thicket, under which it will quickly dodge out of sight at the merest suspicion of a footstep. To most people the bird is only a voice calling, " *TEACHER, TEACHER, TEACHER, TEACHER, TEACHER!*" as Mr. Burroughs has interpreted the notes that go off in pairs like a series of little explosions, softly at first, then louder and louder and more shrill until the bird that you at first thought far away seems to be shrieking his penetrating crescendo into your very ears. But you may look until you are tired before you find him in the high, dry wood, never near water.

In the driest parts of the wood, where the ground is thickly carpeted with dead leaves, you may some day notice a little bunch of them, that look as if a plant, in pushing its way up through the ground, had raised the leaves, rootlets, and twigs a trifle.

Examine the spot more carefully, and on one side you find an opening, and within the ball of earth, softly lined with grass, lie four or five cream-white, speckled eggs. It is only by a happy accident that this nest of the ovenbird is discovered. The concealment could not be better. It is this peculiarity of nest construction—in shape like a Dutch oven—that has given the bird what DeKay considers its "trivial name." Not far from the nest the parent birds scratch about in the leaves, like diminutive barnyard fowls, for the grubs and insects hiding under them. But at the first suspicion of an intruder their alarm becomes pitiful. Panic-stricken, they become fairly limp with fear, and drooping her wings and tail, the mother-bird drags herself hither and thither over the ground.

As utterly bewildered as his mate, the male darts, flies, and tumbles about through the low branches, jerking and wagging his tail in nervous spasms until you have beaten a double-quick retreat.

In nesting time, at evening, a very few have heard the "luxurious nuptial song" of the ovenbird; but it is a song to haunt the memory forever afterward. Burroughs appears to be the first writer to record this "rare bit of bird melody." "Mounting by easy flight to the top of the tallest tree," says the author of "Wake-Robin," "the ovenbird launches into the air with a sort of suspended, hovering flight, like certain of the finches, and bursts into a perfect ecstasy of song—clear, ringing, copious, rivalling the goldfinch's in vivacity and the linnet's in melody."

Worm-eating Warbler

(*Helmintberus vermivorus*) Wood Warbler family

Length—5.50 inches. Less than an inch shorter than the English sparrow.

Male and Female—Greenish olive above. Head yellowish brown, with two black stripes through crown to the nape; also black lines from the eyes to neck. Under parts buffy and white.

Range—Eastern parts of United States. Nests as far north as southern Illinois and southern Connecticut. Winters in the Gulf States and southward.

Migrations—May. September. Summer resident.

In the Delaware Valley and along the same parallel, this inconspicuous warbler is abundant, but north of New Jersey it is rare enough to give an excitement to the day on which you discover it. No doubt it is commoner than we suppose, for its coloring blends so admirably with its habitats that it is probably very often overlooked. Its call-note, a common chirp, has nothing distinguishing about it, and all ornithologists confess to having been often misled by its song into thinking it came from the chipping sparrow. It closely resembles that of the pine warbler also. If it were as nervously active as most warblers, we should more often discover it, but it is quite as deliberate as a vireo, and in the painstaking way in which it often circles around a tree while searching for spiders and other insects that infest the trunks, it reminds us of the brown creeper. Sunny slopes and hillsides covered with thick undergrowth are its preferred foraging and nesting haunts. It is often seen hopping directly on the dry ground, where it places its nest, and it never mounts far above it. The well-drained, sunny situation for the home is chosen with the wisdom of a sanitary expert.

Acadian Flycatcher

(Empidonax virescens) Flycatcher family

Called also: SMALL GREEN-CRESTED FLYCATCHER ; SMALL PEWEE

Length—5.75 to 6 inches. A trifle smaller than the English sparrow.

Male—Dull olive above. Two conspicuous yellowish wing-bars. Throat white, shading into pale yellow on breast. Light gray or white underneath. Upper part of bill black; lower mandible flesh-color. White eye-ring.

Female—Greener above and more yellow below.

Range—From Canada to Mexico, Central America, and West Indies. Most common in south temperate latitudes. Winters in southerly limit of range.

Migrations—April. September Summer resident.

When all our northern landscape takes on the exquisite, soft green, gray, and yellow tints of early spring, this little flycatcher, in perfect color-harmony with the woods it darts among, comes out of the south. It might be a leaf that is being blown about, touched by the sunshine filtering through the trees, and partly shaded by the young foliage casting its first shadows.

Woodlands, through which small streams meander lazily, inviting swarms of insects to their boggy shores, make ideal hunting grounds for the Acadian flycatcher. It chooses a low rather than a high, conspicuous perch, that other members of its family invariably select; and from such a lookout it may be seen launching into the air after the passing gnat—darting downward, then suddenly mounting upward in its aërial hunt, the vigorous clicks of the beak as it closes over its tiny victims testifying to the bird's unerring aim and its hearty appetite.

While perching, a constant tail-twitching is kept up; and a faint, fretful "*Tshee-kee, tshee-kee*" escapes the bird when inactively waiting for a dinner to heave in sight.

In the Middle Atlantic States its peeping sound and the clicking of its particolored bill are infrequently heard in the village streets in the autumn, when the shy and solitary birds are enticed from the deep woods by a prospect of a more plentiful diet of insects, attracted by the fruit in orchards and gardens.

Never far from the ground, on two or more parallel branches, the shallow, unsubstantial nest is laid. Some one has cleverly described it as "a tuft of hay caught by the limb from a load driven under it," but this description omits all mention of the quantities of blossoms that must be gathered to line the cradle for the tiny, pure white eggs.

Yellow-bellied Flycatcher

(*Empidonax flaviventris*) Flycatcher family

Length—5 to 5.6 inches. About an inch smaller than the English sparrow.

Male—Rather dark, but true olive-green above. Throat and breast yellowish olive, shading into pale yellow underneath, including wing linings and under tail coverts. Wings have yellowish bars. Whitish ring around eye. Upper part of bill black, under part whitish or flesh-colored.

Female—Smaller, with brighter yellow under parts and more decidedly yellow wing-bars.

Range—North America, from Labrador to Panama, and westward from the Atlantic to the plains. Winters in Central America.

Migrations—May. September. Summer resident. More commonly a migrant only.

This is the most yellow of the small flycatchers and the only Eastern species with a yellow instead of a white throat. Without hearing its call-note, "*pse-ek-pse-ek*," which it abruptly sneezes rather than utters, it is quite impossible, as it darts among the trees, to tell it from the Acadian flycatcher, with which even Audubon confounded it. Both these little birds choose the same sort of retreats—well-timbered woods near a stream that attracts myriads of insects to its spongy shores—and both are rather shy and solitary. The yellow-bellied species has a far more northerly range, however, than its Southern relative or even the small green-crested flycatcher. It is rare in the Middle States, not common even in New England, except in the migrations, but from the Canada border northward its soft, plaintive whistle, which is its love-song, may be heard in every forest where it nests. All the flycatchers seem to make a noise with so much struggle, such convulsive jerkings of head and tail, and flutterings of the wings that, considering the scanty success of their musical attempts, it is surprising they try to lift their voices at all when the effort almost literally lifts them off their feet.

While this little flycatcher is no less erratic than its Acadian cousin, its nest is never slovenly. One couple had their home in a wild-grape bower in Pennsylvania ; a Virginia creeper in New Jersey supported another cradle that was fully twenty feet above the ground ; but in Labrador, where the bird has its chosen breeding grounds, the bulky nest is said to be invariably placed either in the moss by the brookside or in some old stump, should the locality be too swampy.

Black-throated Green Warbler

(Dendroica virens) Wood Warbler family

Length—5 inches. Over an inch smaller than the English sparrow.
Male—Back and crown of head bright yellowish olive-green. Forehead, band over eye, cheeks, and sides of neck rich yellow. Throat, upper breast, and stripe along sides black. Underneath yellowish white. Wings and tail brownish olive, the former with two white bars, the latter with much

white in outer quills. In autumn, plumage resembling the female's.

Female—Similar ; chin yellowish , throat and breast dusky, the black being mixed with yellowish.

Range—Eastern North America, from Hudson Bay to Central America and Mexico. Nests north of Illinois and New York. Winters in tropics.

Migrations—May. October. Common summer resident north of New Jersey.

There can be little difficulty in naming a bird so brilliantly and distinctly marked as this green, gold, and black warbler, that lifts up a few pure, sweet, tender notes, loud enough to attract attention when he visits the garden. *"See-see, see-saw,"* he sings, but there is a tone of anxiety betrayed in the simple, sylvan strain that always seems as if the bird needed reassuring, possibly due to the rising inflection, like an interrogative, of the last notes.

However abundant about our homes during the migrations, this warbler, true to the family instinct, retreats to the woods to nest—not always so far away as Canada, the nesting ground of most warblers, for in many Northern States the bird is commonly found throughout the summer. Doubtless it prefers tall evergreen trees for its mossy, grassy nest; but it is not always particular, so that the tree be a tall one with a convenient fork in an upper branch.

Early in September increased numbers emerge from the woods, the plumage of the male being less brilliant than when we saw it last, as if the family cares of the summer had proved too taxing. For nearly a month longer they hunt incessantly, with much flitting about the leaves and twigs at the ends of branches in the shrubbery and evergreens, for the tiny insects that the warblers must devour by the million during their all too brief visit.

BIRDS CONSPICUOUSLY YELLOW
AND ORANGE

Yellow-throated Vireo

American Goldfinch

Evening Grosbeak

Blue-winged Warbler

Canadian Warbler

Hooded Warbler

Kentucky Warbler

Magnolia Warbler

Mourning Warbler

Nashville Warbler

Pine Warbler

Prairie Warbler

Wilson's Warbler or Blackcap

Yellow Warbler or Summer Yellowbird

Yellow Redpoll Warbler

Yellow-breasted Chat

Maryland Yellowthroat

Blackburnian Warbler

Redstart

Baltimore Oriole

Look also among the Yellowish Olive Birds in the preceding group; and among the Brown Birds for the Meadowlark and Flicker. See also Parula Warbler (Slate) and Yellow-bellied Woodpecker (Black and White).

YELLOW-THROATED VIREO
Life-size.

BIRDS CONSPICUOUSLY YELLOW AND ORANGE

Yellow-throated Vireo

(Vireo flavifrons) Vireo or Greenlet family

Length—5.5 to 6 inches. A little smaller than the English sparrow.

Male and Female—Lemon-yellow on throat, upper breast; line around the eye and forehead. Yellow, shading into olive-green, on head, back, and shoulders. Underneath white. Tail dark brownish, edged with white. Wings a lighter shade, with two white bands across, and some quills edged with white.

Range—North America, from Newfoundland to Gulf of Mexico, and westward to the Rockies. Winters in the tropics.

Migrations—May. September. Spring and autumn migrant; more rarely resident.

This is undoubtedly the beauty of the vireo family—a group of neat, active, stoutly built, and vigorous little birds of yellow, greenish, and white plumage; birds that love the trees, and whose feathers reflect the coloring of the leaves they hide, hunt, and nest among. "We have no birds," says Bradford Torrey, "so unsparing of their music: they sing from morning till night."

The yellow-throated vireo partakes of all the family characteristics, but, in addition to these, it eclipses all its relatives in the brilliancy of its coloring and in the art of nest-building, which it has brought to a state of hopeless perfection. No envious bird need try to excel the exquisite finish of its workmanship. Happily, it has wit enough to build its pensile nest high above the reach of small boys, usually suspending it from a branch overhanging running water that threatens too precipitous a bath to tempt the young climbers.

However common in the city parks and suburban gardens this bird may be during the migrations, it delights in a secluded retreat overgrown with tall trees and near a stream, such as is dear to the solitary vireo as well when the nesting time approaches. High up in the trees we hear its rather sad, persistent strain, that is more in harmony with the dim forest than with the gay flower garden, where, if the truth must be told, its song is both monotonous and depressing. Mr. Bicknell says it is the only vireo that sings as it flies.

American Goldfinch

(Spinus tristis) Finch family

Called also: WILD CANARY; YELLOWBIRD; THISTLE BIRD

Length—5 to 5.2 inches. About an inch smaller than the English sparrow.

Male—In summer plumage Bright yellow, except on crown of head, frontlet, wings, and tail, which are black. Whitish markings on wings give effect of bands. Tail with white on inner webs. *In winter plumage:* Head yellow-olive; no frontlet; back drab, with reddish tinge; shoulders and throat yellow; soiled brownish white underneath.

Female—Brownish olive above, yellowish white beneath.

Range—North America, from the tropics to the Fur Countries and westward to the Columbia River and California. Common throughout its range.

Migrations—May. October. Common summer resident, frequently seen throughout the winter as well.

An old field, overgrown with thistles and tall, stalky wild flowers, is the paradise of the goldfinches, summer or winter. Here they congregate in happy companies while the sunshine and goldenrod are as bright as their feathers, and cling to the swaying, slender stems that furnish an abundant harvest, daintily lunching upon the fluffy seeds of thistle blossoms, pecking at the mullein-stalks, and swinging airily among the asters and Michaelmas daisies; or, when snow covers the same field with a glistening crust, above which the brown stalks offer only a meagre dinner, the same birds, now sombrely clad in winter feathers, cling to the swaying stems with cheerful fortitude.

At your approach, the busy company rises on the wing, and with peculiar, wavy flight rise and fall through the air, marking each undulation with a cluster of notes, sweet and clear, that come floating downward from the blue ether, where the birds seem to bound along exultant in their motion and song alike.

In the spring the plumage of the goldfinch, which has been drab and brown through the winter months, is moulted or shed—a change that transforms the bird from a sombre Puritan into the gayest of cavaliers, and seems to wonderfully exalt his spirits. He bursts into a wild, sweet, incoherent melody that might be the outpouring from two or three throats at once instead of one, expressing his rapture somewhat after the manner of the canary, although his song lacks the variety and the finish of his caged namesake. What tone of sadness in his music the man found who applied the adjective *tristis* to his scientific name it is difficult to imagine when listening to the notes that come bubbling up from the bird's happy heart.

With plumage so lovely and song so delicious and dreamy, it is small wonder that numbers of our goldfinches are caught and caged, however inferior their song may be to the European species recently introduced into this country. Heard in Central Park, New York, where they were set at liberty, the European goldfinches seemed to sing with more abandon, perhaps, but with no more sweetness than their American cousins. The song remains at its best all through the summer months, for the bird is a long wooer. It is nearly July before he mates, and not until the tardy cedar birds are house-building in the orchard do the happy pair begin to carry grass, moss, and plant-down to a crotch of some tall tree convenient to a field of such wild flowers as will furnish food to a growing family. Doubtless the birds wait for this food to be in proper condition before they undertake parental duties at all—the most plausible excuse for their late nesting. The cares evolving from four to six pale-blue eggs will suffice to quiet the father's song for the winter by the first of September, and fade all the glory out of his shining coat. As pretty a sight as any garden offers is when a family of goldfinches alights on the top of a sunflower to feast upon the oily seeds—a perfect harmony of brown and gold.

Evening Grosbeak

(Coccothraustes vespertinus) Finch family

Length—8 inches. Two inches shorter than the robin.

Male—Forehead, shoulders, and underneath clear yellow; dull yellow on lower back; sides of the head, throat, and breast olive-brown. Crown, tail, and wings black, the latter with white secondary feathers. Bill heavy and blunt, and yellow.

Female—Brownish gray, more or less suffused with yellow Wings and tail blackish, with some white feathers.

Range—Interior of North America. Resident from Manitoba northward. Common winter visitor in northwestern United States and Mississippi Valley; casual winter visitor in northern Atlantic States.

In the winter of 1889-90 Eastern people had the rare treat of becoming acquainted with this common bird of the Northwest, that, in one of its erratic travels, chose to visit New England and the Atlantic States, as far south as Delaware, in great numbers. Those who saw the evening grosbeaks then remember how beautiful their yellow plumage—a rare winter tint—looked in the snow-covered trees, where small companies of the gentle and even tame visitors enjoyed the buds and seeds of the maples, elders, and evergreens. Possibly evening grosbeaks were in vogue for the next season's millinery, or perhaps Eastern ornithologists had a sudden zeal to investigate their structural anatomy. At any rate, these birds, whose very tameness, that showed slight acquaintance with mankind, should have touched the coldest heart, received the warmest kind of a reception from hot shot. The few birds that escaped to the solitudes of Manitoba could not be expected to tempt other travellers eastward by an account of their visit. The bird is quite likely to remain rare in the East.

But in the Mississippi Valley and throughout the northwest, companies of from six to sixty may be regularly counted upon as winter neighbors on almost every farm. Here the females keep up a busy chatting, like a company of cedar birds, and the males punctuate their pauses with a single shrill note that gives little indication of their vocal powers. But in the solitude of the northern forests the love-song is said to resemble the robin's at the start. Unhappily, after a most promising beginning, the bird suddenly stops, as if he were out of breath.

EVENING GROSBEAK
¼ Life size.

Blue-winged Warbler

(Helminthophila pinus) Wood Warbler family

Called also : BLUE-WINGED YELLOW WARBLER

Length—4.75 inches. An inch and a half shorter than the English sparrow.

Male—Crown of head and all under parts bright yellow. Back olive-green. Wings and tail bluish slate, the former with white bars, and three outer tail quills with large white patches on their inner webs.

Female—Paler and more olive.

Range—Eastern United States, from southern New England and Minnesota, the northern limit of its nesting range, to Mexico and Central America, where it winters.

Migrations—May. September. Summer resident.

In the naming of warblers, bluish slate is the shade intended when blue is mentioned; so that if you see a dainty little olive and yellow bird with slate-colored wings and tail hunting for spiders in the blossoming orchard or during the early autumn, you will have seen the beautiful blue-winged warbler. It has a rather leisurely way of hunting, unlike the nervous, restless flitting about from twig to twig that is characteristic of many of its many cousins. The search is thorough—bark, stems, blossoms, leaves are inspected for larvæ and spiders, with many pretty motions of head and body. Sometimes, hanging with head downward, the bird suggests a yellow titmouse. After blossom time a pair of these warblers, that have done serviceable work in the orchard in their all too brief stay, hurry off to dense woods to nest. They are usually to be seen in pairs at all seasons. Not to "high coniferous trees in northern forests"—the Mecca of innumerable warblers—but to scrubby, second growth of woodland borders, or lower trees in the heart of the woods, do these dainty birds retreat. There they build the usual warbler nest of twigs, bits of bark, leaves, and grasses, but with this peculiarity: the numerous leaves with which the nest is wrapped all have their stems pointing upward. Mr. Frank Chapman has admirably defined their song as consisting of "two drawled, wheezy notes —*swee-chee*, the first inhaled, the second exhaled."

Canadian Warbler

(Sylvania canadensis) Wood Warbler family

Called also: CANADIAN FLYCATCHER ; SPOTTED CANA-
DIAN WARBLER

Length—5 to 5.6 inches.   About an inch shorter than the English
sparrow.
Male—Immaculate bluish ash above, without marks on wings or
tail; crown spotted with arrow-shaped black marks. Cheeks,
line from bill to eye, and underneath clear yellow. Black
streaks forming a necklace across the breast.
Female—Paler, with necklace indistinct.
Range—North America, from Manitoba and Labrador to tropics.
Migrations—May. September. Summer resident ; most abundant in migrations.

Since about one-third of all the song-birds met with in a
year's rambles are apt to be warblers, the novice cannot devote
his first attention to a better group, confusing though it is by
reason of its size and the repetition of the same colors in so many
bewildering combinations. Monotony, however, is unknown in
the warbler family. Whoever can rightly name every warbler,
male and female, on sight is uniquely accomplished.

The jet necklace worn on this bird's breast is its best mark
of identification. Its form is particularly slender and graceful, as
might be expected in a bird so active, one to whom a hundred
tiny insects barely afford a dinner that must often be caught piecemeal as it flies past. To satisfy its appetite, which cannot but be
dainty in so thoroughly charming a bird, it lives in low, boggy
woods, in such retreats as Wilson's black-capped warbler selects
for a like reason. Neither of these two " flycatcher " warblers
depends altogether on catching insects on the wing; countless
thousands are picked off the under sides of leaves and about the
stems of twigs in true warbler fashion.

The Canadian's song is particularly loud, sweet, and vivacious. It is hazardous for any one without long field practice to
try to name any warbler by its song alone, but possibly this one's
animated music is as characteristic as any.

The nest is built on the ground on a mossy bank or elevated

194

into the root crannies of some large tree, where there is much water in the woods. Bits of bark, dead wood, moss, and fine rootlets, all carefully wrapped with leaves, go to make the pretty cradle. Unhappily, the little Canada warblers are often cheated out of their natural rights, like so many other delightful song-birds, by the greedy interloper that the cowbird deposits in their nest.

Hooded Warbler

(Sylvania mitrata) Wood Warbler family

Length—5 to 5.75 inches. About an inch shorter than the English sparrow.

Male—Head, neck, chin, and throat black like a hood in mature male specimens only. Hood restricted, or altogether wanting in female and young. Upper parts rich olive. Forehead, cheeks, and underneath yellow. Some conspicuous white on tail feathers.

Female—Duller, and with restricted cowl.

Range—United States east of Rockies, and from southern Michigan and southern New England to West Indies and tropical America, where it winters. Very local.

Migrations—May. September. Summer resident.

This beautifully marked, sprightly little warbler might be mistaken in his immaturity for the yellowthroat ; and as it is said to take him nearly three years to grow his hood, with the completed cowl and cape, there is surely sufficient reason here for the despair that often seizes the novice in attempting to distinguish the perplexing warblers. Like its Southern counterpart, the hooded warbler prefers wet woods and low trees rather than high ones, for much of its food consists of insects attracted by the dampness, and many of them must be taken on the wing. Because of its tireless activity the bird's figure is particularly slender and graceful—a trait, too, to which we owe all the glimpses of it we are likely to get throughout the summer. It has a curious habit of spreading its tail, as if it wished you to take special notice of the white spots that adorn it; not flirting it, as the redstart does his more gorgeous one, but simply opening it like a fan as it flies and darts about.

Its song, which is particularly sweet and graceful, and with more variation than most warblers' music, has been translated "*Che-we-eo-tsip, tsip, che-we-eo,*" again interpreted by Mr. Chapman as "You must come to the woods, or you won't see me."

Kentucky Warbler

(Geothlypis formosa) Wood Warbler family

Length—5.5 inches. Nearly an inch shorter than the English sparrow.

Male—Upper parts olive-green; under parts yellow; a yellow line from the bill passes over and around the eye. Crown of head, patch below the eye, and line defining throat, black.

Female—Similar, but paler, and with grayish instead of black markings.

Range—United States eastward from the Rockies, and from Iowa and Connecticut to Central America, where it winters.

Migrations—May. September. Summer resident.

No bird is common at the extreme limits of its range, and so this warbler has a reputation for rarity among the New England ornithologists that would surprise people in the middle South and Southwest. After all that may be said in the books, a bird is either common or rare to the individual who may or may not have happened to become acquainted with it in any part of its chosen territory. Plenty of people in Kentucky, where we might judge from its name this bird is supposed to be most numerous, have never seen or heard of it, while a student on the Hudson River, within sight of New York, knows it intimately. It also nests regularly in certain parts of the Connecticut Valley. "Who is my neighbor?" is often a question difficult indeed to answer where birds are concerned. In the chapter, "Spring at the Capital," which, with every reading of "Wake Robin," inspires the bird-lover with fresh zeal, Mr. Burroughs writes of the Kentucky warbler: "I meet with him in low, damp places, in the woods, usually on the steep sides of some little run. I hear at intervals a clear, strong, bell-like whistle or warble, and presently catch a glimpse of the bird as he jumps up from the ground to take an insect or worm from the under side of a leaf. This is his characteristic movement. He belongs to the class of ground warblers, and his range is very low, indeed lower than that of any other species with which I am acquainted."

Like the ovenbird and comparatively few others, for most birds hop over the ground, the Kentucky warbler *walks* rapidly about, looking for insects under the fallen leaves, and poking his inquisitive beak into every cranny where a spider may be lurking. The bird has a pretty, conscious way of flying up to a perch, a few feet above the ground, as a tenor might advance towards the footlights of a stage, to pour forth his clear, penetrating whistle, that in the nesting season especially is repeated over and over again with tireless persistency.

Magnolia Warbler

(Dendroica maculosa) Wood Warbler family

Called also: BLACK-AND-YELLOW WARBLER ; SPOTTED WARBLER ; BLUE-HEADED YELLOW-RUMPED WARBLER

Length—4.75 to 5 inches. About an inch and a half smaller than the English sparrow.

Male—Crown of head slate-color, bordered on either side by a white line ; a black line, apparently running through the eye, and a yellow line below it, merging into the yellow throat. Lower back and under parts yellow. Back, wings, and tail blackish olive. Large white patch on the wings, and the middle of the tail-quills white. Throat and sides heavily streaked with black.

Female—Has greener back, is paler, and has less distinct markings.

Range—North America, from Hudson Bay to Panama. Summers from northern Michigan and northern New England northward ; winters in Central America and Cuba.

Migrations—May. October. Spring and summer migrant.

In spite of the bird's name, one need not look for it in the glossy magnolia trees of the southern gardens more than in the shrubbery on New England lawns, and during the migrations it is quite as likely to be found in one place as in the other. Its true preference, however, is for the spruces and hemlocks of its nesting ground in the northern forests. For these it deserts us after a brief hunt about the tender, young spring foliage and blossoms, where the early worm lies concealed, and before we have become so well acquainted with its handsome clothes that we will instantly recognize it in the duller ones it wears on its return

197

trip in the autumn. The position of the white marks on the tail feathers of this warbler, however, is the clue by which it may be identified at any season or any stage of its growth. If the white bar runs across the *middle* of the warbler's tail, you can be sure of the identity of the bird. A nervous and restless hunter, it nevertheless seems less shy than many of its kin. Another pleasing characteristic is that it brings back with it in October the loud, clear, rapid whistle with which it has entertained its nesting mate in the Canada woods through the summer.

Mourning Warbler

(Geothlypis philadelphia) Wood Warbler family

Called also : MOURNING GROUND WARBLER

Length—5 to 5.6 inches. About an inch smaller than the English sparrow.

Male—Gray head and throat; the breast gray; the feathers with black edges that make them look crinkled, like crape. The black markings converge into a spot on upper breast. Upper parts, except head, olive. Underneath rich yellow.

Female—Similar, but duller; throat and breast buff and dusky where the male is black. Back olive-green.

Range—"Eastern North America; breeds from eastern Nebraska, northern New York, and Nova Scotia northward, and southward along the Alleghanies to Pennsylvania. Winters in the tropics."—*Chapman.*

Migrations—May. September. Spring and autumn migrant.

Since Audubon met with but one of these birds in his incessant trampings, and Wilson secured only an immature, imperfectly marked specimen for his collection, the novice may feel no disappointment if he fails to make the acquaintance of this "gay and agreeable widow." And yet the shy and wary bird is not unknown in Central Park, New York City. Even where its clear, whistled song strikes the ear with a startling novelty that invites to instant pursuit of the singer, you may look long and diligently through the undergrowth without finding it. Dr. Merriam says the whistle resembles the syllables "*true, true, true, tru, too,* the voice rising on the first three syllables and falling on the last two." In the nesting season this song is

198

repeated over and over again with a persistency worthy of a Kentucky warbler. It is delivered from a perch within a few feet of the ground, as high as the bird seems ever inclined to ascend.

Nashville Warbler

(Helminthophila ruficapilla) Wood Warbler family

Length—4.75 to 5 inches. About an inch and a half smaller than the English sparrow.
Male—Olive-green above; yellow underneath. Slate-gray head and neck. Partially concealed chestnut patch on crown. Wings and tail olive-brown and without markings.
Female—Dull olive and paler, with brownish wash underneath.
Range—North America, westward to the plains; north to the Fur Countries, and south to Central America and Mexico. Nests north of Illinois and northern New England; winters in tropics.
Migrations—April. September or October.

It must not be thought that this beautiful warbler confines itself to backyards in the city of Nashville simply because Wilson discovered it near there and gave it a local name, for the bird's actual range reaches from the fur trader's camp near Hudson Bay to the adobe villages of Mexico and Central America, and over two thousand miles east and west in the United States. It chooses open rather than dense woods and tree-bordered fields. It seems to have a liking for hemlocks and pine trees, especially if near a stream that attracts insects to its shores ; and Dr. Warren notes that in Pennsylvania he finds small flocks of these warblers in the autumn migration, feeding in the willow trees near little rivers and ponds. Only in the northern parts of the United States is their nest ever found, for the northern British provinces are their preferred nesting ground. One seen in the White Mountains was built on a mossy, rocky ledge, directly on the ground at the foot of a pine tree, and made of rootlets, moss, needles from the trees overhead, and several layers of leaves outside, with a lining of fine grasses that cradled four white, speckled eggs.

Audubon likened the bird's feeble note to the breaking of twigs.

Pine Warbler

(Dendroica vigorsii) Wood Warbler family

Called also : PINE-CREEPING WARBLER

Length—5.5 to 6 inches. A trifle smaller than the English sparrow.

Male—Yellowish olive above; clear yellow below, shading to grayish white, with obscure dark streaks on side of breast. Two whitish wing-bars; two outer tail feathers partly white.

Female—Duller , grayish white only faintly tinged with yellow underneath.

Range—North America, east of the Rockies; north to Manitoba, and south to Florida and the Bahamas. Winters from southern Illinois southward.

Migrations—March or April. October or later Common summer resident.

The pine warbler closely presses the myrtle warbler for the first place in the ranks of the family migrants, but as the latter bird often stays north all winter, it is usually given the palm. Here is a warbler, let it be recorded, that is fittingly named, for it is a denizen of pine woods only; most common in the long stretches of pine forests at the south and in New York and New England, and correspondingly uncommon wherever the woodsman's axe has laid the pine trees low throughout its range. Its "simple, sweet, and drowsy song," writes Mr. Parkhurst, is always associated "with the smell of pines on a sultry day." It recalls that of the junco and the social sparrow or chippy.

Creeping over the bark of trees and peering into every crevice like a nuthatch; running along the limbs, not often hopping nervously or flitting like the warblers; darting into the air for a passing insect, or descending to the ground to feed on seeds and berries, the pine warbler has, by a curious combination, the movements that seem to characterize several different birds.

It is one of the largest and hardiest members of its family, but not remarkable for its beauty. It is a sociable traveller, cheerfully escorting other warblers northward, and welcoming to its band both the yellow redpolls and the myrtle warblers. These birds are very often seen together in the pine and other evergreen trees in our lawns and in the large city parks.

Prairie Warbler

(Dendroica discolor) Wood Warbler family

Length—4.75 to 5 inches. About an inch and a half shorter than the English sparrow.

Male—Olive-green above, shading to yellowish on the head, and with brick-red spots on back between the shoulders. A yellow line over the eye; wing-bars and all under parts bright yellow, heavily streaked with black on the sides. Line through the eye and crescent below it, black. Much white in outer tail feathers.

Female—Paler; upper parts more grayish olive, and markings less distinct than male's.

Range—Eastern half of the United States. Nests as far north as New England and Michigan. Winters from Florida southward.

Migrations—May. September. Summer resident.

Doubtless this diminutive bird was given its name because it prefers open country rather than the woods—the scrubby undergrowth of oaks, young evergreens, and bushes that border clearings being as good a place as any to look for it, and not the wind-swept, treeless tracts of the wild West. Its range is southerly. The Southern and Middle States are where it is most abundant. Here is a wood warbler that is not a bird of the woods—less so, in fact, than either the summer yellowbird (yellow warbler) or the palm warbler, that are eminently neighborly and fond of pasture lands and roadside thickets. But the prairie warblers are rather more retiring little sprites than their cousins, and it is not often we get a close enough view of them to note the brick-red spots on their backs, which are their distinguishing marks. They have a most unkind preference for briery bushes, that discourage human intimacy. In such forbidding retreats they build their nest of plant-fibre, rootlets, and twigs, lined with plant-down and hair.

The song of an individual prairie warbler makes only a slight impression. It consists "of a series of six or seven quickly repeated *zees*, the next to the last one being the highest" (Chapman). But the united voices of a dozen or more of these pretty little birds, that often sing together, afford something approaching a musical treat.

Wilson's Warbler

(Sylvania pusilla) Wood Warbler family

Called also : BLACKCAP ; GREEN BLACK-CAPPED WAR-
BLER ; WILSON'S FLYCATCHER

Length—4.75 to 5 inches. About an inch and a half shorter than
the English sparrow
Male—Black cap ; yellow forehead ; all other upper parts olive-
green ; rich yellow underneath.
Female—Lacks the black cap.
Range—North America, from Alaska and Nova Scotia to Panama.
Winters south of Gulf States. Nests chiefly north of the
United States.
Migrations—May. September. Spring and autumn migrant.

To see this strikingly marked little bird one must be on the
sharp lookout for it during the latter half of May, or at the season
of apple bloom, and the early part of September. It passes northward with an almost scornful rapidity. Audubon mentions having seen it in Maine at the end of October, but this specimen
surely must have been an exceptional laggard.

In common with several others of its family, it is exceedingly
expert in catching insects on the wing ; but it may be known as
no true flycatcher from the conspicuous rich yellow of its under
parts, and also from its habit of returning from a midair sally to a
different perch from the one it left to pursue its dinner. A true
flycatcher usually returns to its old perch after each hunt.

To indulge in this aërial chase with success, these warblers
select for their home and hunting ground some low woodland
growth where a sluggish stream attracts myriads of insects to
the boggy neighborhood. Here they build their nest in low
bushes or upon the ground. Four or five grayish eggs, sprinkled
with cinnamon-colored spots in a circle around the larger end,
are laid in the grassy cradle in June. Mr. H. D. Minot found one
of these nests on Pike's Peak at an altitude of 11,000 feet, almost
at the limit of vegetation. The same authority compares the
bird's song to that of the redstart and the yellow warbler.

Yellow Redpoll Warbler

(Dendroica palmarum hypochrysea) Wood Warbler family

Called also : YELLOW PALM WARBLER

Length—5.5 to 5.75 inches. A little smaller than the English sparrow.

Male and Female—Chestnut crown. Upper parts brownish olive ; greenest on lower back. Underneath uniform bright yellow, streaked with chestnut on throat, breast, and sides. Yellow line over and around the eye. Wings unmarked. Tail edged with olive-green; a few white spots near tips of outer quills. More brownish above in autumn, and with a grayish wash over the yellow under parts.

Range—Eastern parts of North America. Nests from Nova Scotia northward. Winters in the Gulf States.

Migrations—April. October. Spring and autumn migrant.

While the uniform yellow of this warbler's under parts in any plumage is its distinguishing mark, it also has a flycatcher's trait of constantly flirting its tail, that is at once an outlet for its superabundant vivacity and a fairly reliable aid to identification. The tail is jerked, wagged, and flirted like a baton in the hands of an inexperienced leader of an orchestra. One need not go to the woods to look for the restless little sprite that comes northward when the early April foliage is as yellow and green as its feathers. It prefers the fields and roadsides, and before there are leaves enough on the undergrowth to conceal it we may come to know it as well as it is possible to know any bird whose home life is passed so far away. Usually it is the first warbler one sees in the spring in New York and New England. With all the alertness of a flycatcher, it will dart into the air after insects that fly near the ground, keeping up a constant *chip, chip,* fine and shrill, at one end of the small body, and the liveliest sort of tail motions at the other. The pine warbler often bears it company.

With the first suspicion of warm weather, off goes this hardy little fellow that apparently loves the cold almost well enough to stay north all the year like its cousin, the myrtle warbler. It builds a particularly deep nest, of the usual warbler construction, on the ground, but its eggs are rosy rather than the bluish white of others.

In the Southern States the bird becomes particularly neighborly, and is said to enter the streets and gardens of towns with a chippy's familiarity.

Palm Warbler or Redpoll Warbler *(Dendroica palmarum)* differs from the preceding chiefly in its slightly smaller size, the more grayish-brown tint in its olive upper parts, and the uneven shade of yellow underneath that varies from clear yellow to soiled whitish. It is the Western counterpart of the yellow redpoll, and is most common in the Mississippi Valley. Strangely enough, however, it is this warbler, and not *hypochrysea*, that goes out of its way to winter in Florida, where it is abundant all winter.

Yellow Warbler

(Dendroica æstiva) Wood Warbler family

Called also : SUMMER YELLOWBIRD ; GOLDEN WARBLER ; YELLOW POLL

Length—4.75 to 5.2 inches. Over an inch shorter than the English sparrow.

Male—Upper parts olive-yellow, brightest on the crown; under parts bright yellow, streaked with reddish brown. Wings and tail dusky olive-brown, edged with yellow.

Female—Similar; but reddish-brown streakings less distinct.

Range—North America, except Southwestern States, where the prothonotary warbler reigns in its stead. Nests from Gulf States to Fur Countries. Winters south of the Gulf States, as far as northern parts of South America.

Migrations—May. September. Common summer resident.

This exquisite little creature of perpetual summer (though to find it it must travel back and forth between two continents) comes out of the south with the golden days of spring. From much living in the sunshine through countless generations, its feathers have finally become the color of sunshine itself, and in disposition, as well, it is nothing if not sunny and bright. Not the least of its attractions is that it is exceedingly common everywhere: in the shrubbery of our lawns, in gardens and orchards, by the road and brookside, in the edges of woods—everywhere we catch its glint of brightness through the long summer days, and hear its simple, sweet, and happy song until the end of July.

SUMMER YELLOW-BIRD.
⅗ Life-size.

Because both birds are so conspicuously yellow, no doubt this warbler is quite generally confused with the goldfinch; but their distinctions are clear enough to any but the most superficial glance. In the first place, the yellow warbler is a smaller bird than the goldfinch; it has neither black crown, wings, nor tail, and it does have reddish-brown streaks on its breast that are sufficiently obsolete to make the coloring of that part look simply dull at a little distance. The goldfinch's bill is heavy, in order that it may crack seeds, whereas the yellow warbler's is slender, to enable it to pick minute insects from the foliage. The goldfinch's wavy, curved flight is unique, and that of his "double" differs not a whit from that of all nervous, flitting warblers. Surely no one familiar with the rich, full, canary-like song of the "wild canary," as the goldfinch is called, could confuse it with the mild "Wee-chee, chee, cher-wee" of the summer yellowbird. Another distinction, not always infallible, but nearly so, is that when seen feeding, the goldfinch is generally below the line of vision, while the yellow warbler is either on it or not far above it, as it rarely goes over twelve feet from the ground.

No doubt, the particularly mild, sweet amiability of the yellow warbler is responsible for the persistent visitations of the cowbird, from which it is a conspicuous sufferer. In the exquisite, neat little matted cradle of glistening milk-weed flax, lined with down from the fronds of fern, the skulking housebreaker deposits her surreptitious egg for the little yellow mother-bird to hatch and tend. But amiability is not the only prominent trait in the female yellow warbler's character. She is clever as well, and quickly builds a new bottom on her nest, thus sealing up the cowbird's egg, and depositing her own on the soft, spongy floor above it. This operation has been known to be twice repeated, until the nest became three stories high, when a persistent cowbird made such unusual architecture necessary.

The most common nesting place of the yellow warbler is in low willows along the shores of streams.

Yellow-breasted Chat

(Icteria virens) Wood Warbler family

Called also: POLYGLOT CHAT ; YELLOW MOCKING-BIRD

Length—7.5 inches. A trifle over an inch longer than the English sparrow.

Male and Female—Uniform olive-green above. Throat, breast, and under side of wings bright, clear yellow. Underneath white. Sides grayish. White line over the eye, reaching to base of bill and forming partial eye-ring. Also white line on sides of throat. Bill and feet black.

Range—North America, from Ontario to Central America and westward to the plains. Most common in Middle Atlantic States.

Migrations—Early May. Late August or September. Summer resident.

This largest of the warblers might be mistaken for a dozen birds collectively in as many minutes; but when it is known that the jumble of whistles, parts of songs, chuckles, clucks, barks, quacks, whines, and wails proceed from a single throat, the yellow-breasted chat becomes a marked specimen forthwith—a conspicuous individual never to be confused with any other member of the feathered tribe. It is indeed absolutely unique. The catbird and the mocking-bird are rare mimics; but while the chat is not their equal in this respect, it has a large repertoire of weird, uncanny cries all its own—a power of throwing its voice, like a human ventriloquist, into unexpected corners of the thicket or meadow. In addition to its extraordinary vocal feats, it can turn somersaults and do other clown-like stunts as well as any variety actor on the Bowery stage.

Only by creeping cautiously towards the roadside tangle, where this "rollicking polyglot" is entertaining himself and his mate, brooding over her speckled eggs in a bulky nest set in a most inaccessible briery part of the thicket, can you hope to hear him rattle through his variety performance. Walk boldly or noisily past his retreat, and there is "silence there and nothing more." But two very bright eyes peer out at you through the undergrowth, where the trim, elegant-looking bird watches you with quizzical suspicion until you quietly seat yourself and

YELLOW-BREASTED CHAT
½ Life-size.

assume silent indifference. *"Whew, whew!"* he begins, and then immediately, with evident intent to amuse, he rattles off an indescribable, eccentric medley until your ears are tired listening. With bill uplifted, tail drooping, wings fluttering at his side, he cuts an absurd figure enough, but not so comical as when he rises into the air, trailing his legs behind him stork-fashion. This surely is the clown among birds. But zany though he is, he is as capable of devotion to his Columbine as Punchinello, and remains faithfully mated year after year. However much of a tease and a deceiver he may be to the passer-by along the roadside, in the privacy of the domestic circle he shows truly lovable traits.

He has the habit of singing in his unmusical way on moonlight nights. Probably his ventriloquial powers are cultivated not for popular entertainment, but to lure intruders away from his nest.

Maryland Yellowthroat

(Geothlypis trichas) Wood Warbler family

Called also: BLACK-MASKED GROUND WARBLER

Length—5.33 inches. Just an inch shorter than the typical English sparrow.

Male—Olive-gray on head, shading to olive-green on all the other upper parts. Forehead, cheeks, and sides of head black, like a mask, and bordered behind by a grayish line. Throat and breast bright yellow, growing steadily paler underneath.

Female—Either totally lacks black mask or its place is indicated by only a dusky tint. She is smaller and duller.

Range—Eastern North America, west to the plains; most common east of the Alleghanies. Nests from the Gulf States to Labrador and Manitoba; winters south of Gulf States to Panama.

Migrations—May. September. Common summer resident.

"Given a piece of marshy ground with an abundance of skunk cabbage and a fairly dense growth of saplings, and near by a tangle of green brier and blackberry, and you will be pretty sure to have it tenanted by a pair of yellowthroats," says Dr. Abbott, who found several of their nests in skunk-cabbage plants, which he says are favorite cradles. No animal cares to touch this plant if it can be avoided; but have the birds themselves no sense of smell?

Before and after the nesting season these active birds, plump of form, elegant of attire, forceful, but not bold, enter the scrubby pastures near our houses and the shrubbery of old-fashioned, overgrown gardens, and peer out at the human wanderer therein with a charming curiosity. The bright eyes of the male masquerader shine through his black mask, where he intently watches you from the tangle of syringa and snowball bushes ; and as he flies into the laburnum with its golden chain of blossoms that pale before the yellow of his throat and breast, you are so impressed with his grace and elegance that you follow too audaciously, he thinks, and off he goes. And yet this is a bird that seems to delight in being pursued. It never goes so far away that you are not tempted to follow it, though it be through dense undergrowth and swampy thickets, and it always gives you just glimpse enough of its beauties and graces before it flies ahead, to invite the hope of a closer inspection next time. When it dives into the deepest part of the tangle, where you can imagine it hunting about among the roots and fallen leaves for the larvæ, caterpillars, spiders, and other insects on which it feeds, it sometimes amuses itself with a simple little song between the hunts. But the bird's indifference, you feel sure, arises from preoccupation rather than rudeness.

If, however, your visit to the undergrowth is unfortunately timed and there happens to be a bulky nest in process of construction on the ground, a quickly repeated, vigorous *chit, pit, quit,* impatiently inquires the reason for your bold intrusion. Withdraw discreetly and listen to the love-song that is presently poured out to reassure his plain little maskless mate. The music is delivered with all the force and energy of his vigorous nature and penetrates to a surprising distance. " *Follow me, follow me, follow me,*" many people hear him say; others write the syllables, " *Wichity, wichity, wichity, wichity* "; and still others write them, " *I beseech you, I beseech you, I beseech you,*" though the tones of this self-assertive bird rather command than entreat. Mr. Frank Chapman says of the yellowthroats : "They sing throughout the summer, and in August add a flight-song to their repertoire. This is usually uttered toward evening, when the bird springs several feet into the air, hovers for a second, and then drops back to the bushes."

BLACKBURNIAN WARBLER.
Life-size.

Blackburnian Warbler

(Dendroica blackburniæ) Wood Warbler family

Called also: HEMLOCK WARBLER ; ORANGE-THROATED
WARBLER ; TORCH-BIRD

Length—4.5 to 5.5 inches. An inch and a half smaller than the
English sparrow.
Male—Head black, striped with orange-flame ; throat and breast
orange, shading through yellow to white underneath ;
wings, tail, and part of back black, with white markings.
Female—Olive-brown above, shading into yellow on breast, and
paler under parts.
Range—Eastern North America to plains. Winters in tropics.
Migrations—May. September. Spring and autumn migrant.

"The orange-throated warbler would seem to be his right
name, his characteristic cognomen," says John Burroughs, in ever-
delightful "Wake Robin" ; "but no, he is doomed to wear the
name of some discoverer, perhaps the first who robbed his nest
or rifled him of his mate — Blackburn ; hence, Blackburnian
warbler. The *burn* seems appropriate enough, for in these dark
evergreens his throat and breast show like flame. He has a very
fine warble, suggesting that of the redstart, but not especially
musical."

No foliage is dense enough to hide, and no autumnal tint too
brilliant to outshine this luminous little bird that in May, as it
migrates northward to its nesting ground, darts in and out of the
leafy shadows like a tongue of fire.

It is by far the most glorious of all the warblers—a sort of
diminutive oriole. The quiet-colored little mate flits about after
him, apparently lost in admiration of his fine feathers and the
ease with which his thin tenor voice can end his lover's warble in
a high Z.

Take a good look at this attractive couple, for in May they
leave us to build a nest of bark and moss in the evergreens of
Canada—that paradise for warblers—or of the Catskills and Adiron-
dacks, and in autumn they hurry south to escape the first frosts.

Redstart

(Setophaga ruticilla) Wood Warbler family

Called also : YELLOW-TAILED WARBLER

Length—5 to 5.5 inches.

Male—In spring plumage: Head, neck, back, and middle breast glossy black, with blue reflections. Breast and underneath white, slightly flushed with salmon, increasing to bright salmon-orange on the sides of the body and on the wing linings. Occasional specimens show orange-red. Tail feathers partly black, partly orange, with broad black band across the end. Orange markings on wings. Bill and feet black. *In autumn :* Fading into rusty black, olive, and yellow.

Female—Olive-brown, and yellow where the male is orange. Young browner than the females.

Range—North America to upper Canada. West occasionally, as far as the Pacific coast, but commonly found in summer in the Atlantic and Middle States.

Migrations—Early May. End of September. Summer resident.

Late some evening, early in May, when one by one the birds have withdrawn their voices from the vesper chorus, listen for the lingering " *'tsee, 'tsee,'tseet* " (usually twelve times repeated in a minute), that the redstart sweetly but rather monotonously sings from the evergreens, where, as his tiny body burns in the twilight, Mrs. Wright likens him to a " wind-blown firebrand, half glowing, half charred."

But by daylight this brilliant little warbler is constantly on the alert. It is true he has the habit, like the flycatchers (among which some learned ornithologists still class him), of sitting pensively on a branch, with fluffy feathers and drooping wings; but the very next instant he shows true warbler blood by making a sudden dash upward, then downward through the air, tumbling somersaults, as if blown by the wind, flitting from branch to branch, busily snapping at the tiny insects hidden beneath the leaves, clinging to the tree-trunk like a creeper, and singing between bites.

Possibly he will stop long enough in his mad chase to open and shut his tail, fan-fashion, with a dainty egotism that, in the peacock, becomes rank vanity.

BALTIMORE ORIOLE.
⅔ Life-size.

The Germans call this little bird *roth Stert* (red tail), but, like so many popular names, this is a misnomer, as, strictly speaking, the redstart is never red, though its salmon-orange markings often border on to orange-flame.

In a fork of some tall bush or tree, placed ten or fifteen feet from the ground, a carefully constructed little nest is made of moss, horsehair, and strippings from the bark, against which the nest is built, the better to conceal its location. Four or five whitish eggs, thickly sprinkled with pale brown and lilac, like the other warblers', are too jealously guarded by the little mother-bird to be very often seen.

Baltimore Oriole

(Icterus galbula) Oriole and Blackbird family

Called also : GOLDEN ORIOLE; FIREBIRD; GOLDEN ROBIN; HANG-NEST; ENGLISH ROBIN

Length—7 to 8 inches.   About one-fifth smaller than the robin.
Male—Head, throat, upper part of back glossy black.   Wings black, with white spots and edgings.   Tail-quills black, with yellow markings on the tips.   Everywhere else orange, shading into flame.
Female—Yellowish olive.   Wings dark brown, and quills margined with white.   Tail yellowish brown, with obscure, dusky bars.
Range—The whole United States.   Most numerous in Eastern States below 55° north latitude.
Migrations—Early May.   Middle of September.   Common summer resident.

A flash of fire through the air; a rich, high, whistled song floating in the wake of the feathered meteor: the Baltimore oriole cannot be mistaken.   When the orchards are in blossom he arrives in full plumage and song, and awaits the coming of the female birds, that travel northward more leisurely in flocks.   He is decidedly in evidence.   No foliage is dense enough to hide his brilliancy; his temper, quite as fiery as his feathers, leads him into noisy quarrels, and his insistent song with its martial, interrogative notes becomes almost tiresome until he is happily mated and family cares check his enthusiasm.

Conspicuously Yellow and Orange.

Among the best architects in the world is his plain but energetic mate. Gracefully swung from a high branch of some tall
tree, the nest is woven with exquisite skill into a long, flexible
pouch that rain cannot penetrate, nor wind shake from its horsehair moorings. Bits of string, threads of silk, and sometimes
yarn of the gayest colors, if laid about the shrubbery in the garden,
will be quickly interwoven with the shreds of bark and milk-
weed stalks that the bird has found afield. The shape of the
nest often differs, because in unsettled regions, where hawks
abound, it is necessary to make it deeper than seven inches (the
customary depth when it is built near the homes of men), and to
partly close it at the top to conceal the sitting bird. From four
to six whitish eggs, scrawled over with black-brown, are hatched
by the mother oriole, and most jealously guarded by her now
truly domesticated mate.

The number of grubs, worms, flies, caterpillars, and even
cocoons, that go to satisfy the hunger of a family of orioles in a
day, might indicate, if it could be computed, the great value these
birds are about our homes, aside from the good cheer they bring.

There is a popular tradition about the naming of this gorgeous
bird: When George Calvert, the first Lord Baltimore, worn out
and discouraged by various hardships in his Newfoundland colony,
decided to visit Virginia in 1628, he wrote that nothing in the
Chesapeake country so impressed him as the myriads of birds
in its woods. But the song and color of the oriole particularly
cheered and delighted him, and orange and black became the
heraldic colors of the first lords proprietors of Maryland.

Hush ! 'tis he !

My Oriole, my glance of summer fire,
Is come at last; and ever on the watch,
Twitches the pack-thread I had lightly wound
About the bough to help his housekeeping.
Twitches and scouts by turns, blessing his luck,
Yet fearing me who laid it in his way.
Nor, more than wiser we in our affairs,
Divines the Providence that hides and helps.
Heave, ho ! Heave, ho ! he whistles as the twine
Slackens its hold; *once more, now !* and a flash
Lightens across the sunlight to the elm
Where his mate dangles at her cup of felt.

—*James Russell Lowell.*

BIRDS CONSPICUOUSLY RED OF ANY SHADE

Cardinal Grosbeak

Summer Tanager

Scarlet Tanager

Pine Grosbeak

American Crossbill and the White-winged Crossbill

Redpoll and Greater Redpoll

Purple Finch

Robin

Orchard Oriole

See the Red-winged Blackbird (Black). See also the males of the Rose-breasted Grosbeak, the Woodpeckers, the Chewink (Black and White); the Red-breasted Nuthatch, the Bay-breasted and the Chestnut-sided Warblers (Slate and Gray); the Bluebird and Barn Swallow (Blue); the Flicker (Brown); the Humming-bird and the Kinglets (Greenish Gray); and the Blackburnian and Redstart Warblers, and the Baltimore Oriole (Orange).

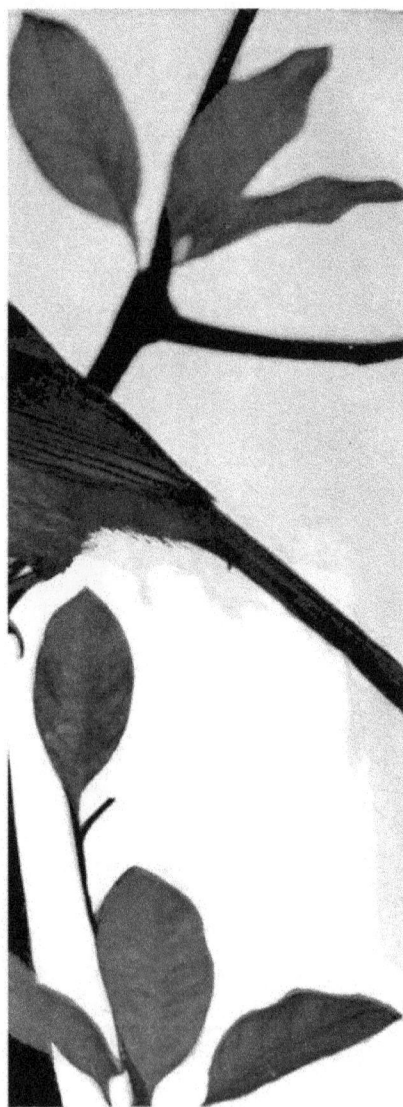

BIRDS CONSPICUOUSLY RED OF ANY SHADE

Cardinal Grosbeak

(Cardinal cardinalis) Finch family

Called also: CRESTED REDBIRD ; VIRGINIA REDBIRD ; VIRGINIA NIGHTINGALE ; CARDINAL BIRD

Length—8 to 9 inches. A little smaller than the robin.

Male—Brilliant cardinal ; chin and band around bill black. Beak stout and red. Crest conspicuous. In winter dress, wings washed with gray.

Female—Brownish yellow above, shading to gray below. Tail shorter than the male's. Crest, wings, and tail reddish. Breast sometimes tinged with red.

Range—Eastern United States. A Southern bird, becoming more and more common during the summer in States north of Virginia, especially in Ohio, south of which it is resident throughout the year.

Migrations—Resident rather than migrating birds, remaining throughout the winter in localities where they have found their way. Travel in flocks.

Among the numerous names by which this beautiful bird is known, it has become immortalized under the title of Mr. James Lane Allen's exquisite book, "The Kentucky Cardinal." Here, while we are given a most charmingly sympathetic, delicate account of the bird "who has only to be seen or heard, and Death adjusts an arrow," it is the cardinal's pathetic fate that impresses one most. Seen through less poetical eyes, however, the bird appears to be a haughty autocrat, a sort of "F. F. V." among the feathered tribes, as, indeed, his title, "Virginia redbird," has been unkindly said to imply. Bearing himself with a refined and courtly dignity, not stooping to soil his feet by walking on the ground like the more democratic robin, or even condescending below the level of the laurel bushes, the cardinal is literally a shining example of self-conscious superiority—a bird to call forth respect and admiration rather than affection. But a group of cardinals in a cedar tree in a snowy winter landscape makes us forgetful of everything but their supreme beauty.

As might be expected in one of the finch family, the cardinal is a songster—the fact which, in connection with his lovely plumage, accounts for the number of these birds shipped in cages to Europe, where they are known as Virginia nightingales. Commencing with a strong, rich whistle, like the high notes of a fife, " *Cheo-cheo-cheo-cheo*," repeated over and over as if to make perfect the start of a song he is about to sing, suddenly he stops, and you learn that there is to be no glorious performance after all, only a prelude to—nothing. The song, such as it is, begins, with both male and female, in March, and lasts, with a brief intermission, until September—"the most melodious sigh," as Mr. Allen calls it. Early in May the cardinals build a bulky and loosely made nest, usually in the holly, laurel, or other evergreen shrubs that they always love to frequent, especially if these are near fields of corn or other grain. And often two broods in a year come forth from the pale-gray, brown-marked eggs, bearing what is literally for them the "fatal gift of beauty."

Summer Tanager

(Piranga rubra) Tanager family

Called also : REDBIRD; SMOOTH-HEADED REDBIRD

Length—7.5 inches. About one-fourth smaller than the robin.
Male—Uniform red. Wings and tail like the body.
Female—Upper parts yellowish olive-green; underneath inclining to orange-yellow.
Range—Tropical portions of two Americas and eastern United States. Most common in Southern States. Rare north of Pennsylvania. Winters in the tropics.
Migrations—In Southern States : April. October Irregular migrant north of the Carolinas.

Thirty years ago, it is recorded that so far north as New Jersey the summer redbird was quite as common as any of the thrushes. In the South still there is scarcely an orchard that does not contain this tropical-looking beauty—the redbird *par excellence*, the sweetest singer of the family. Is there a more beautiful sight in all nature than a grove of orange trees laden with fruit, starred with their delicious blossoms, and with flocks of redbirds disporting themselves among the dark, glossy leaves? Pine and oak woods are also favorite resorts, especially at the north, where the bird nowadays forsakes the orchards to hide his beauty, if he can, unharmed by the rifle that only rarely is offered so shining a mark. He shows the scarlet tanager's preference for tree-tops, where his musical voice, calling "*Chicky-tucky-tuk*," alone betrays his presence in the woods. The Southern farmers declare that he is an infallible weather prophet, his "*WET, WET, WET*," being the certain indication of rain—another absurd saw, for the call-note is by no means confined to the rainy season.

The yellowish-olive mate, whose quiet colors betray no nest secrets, collects twigs and grasses for the cradle to be saddled on the end of some horizontal branch, though in this work the male sometimes cautiously takes an insignificant part. After her three or four eggs are laid she sits upon them for nearly two weeks, being only rarely and stealthily visited by her mate with some choice grub, blossom, or berry in his beak. But how cheerfully his fife-like whistle rings out during the temporary exile! Then his song is at its best. Later in the summer he has an aggravating way of joining in the chorus of other birds' songs, by which the pleasant individuality of his own voice is lost.

A nest of these tanagers, observed not far from New York City, was commenced the last week of May on the extreme edge of a hickory limb in an open wood; four eggs were laid on the fourth of June, and twelve days later the tiny fledglings, that all look like their mother in the early stages of their existence, burst from the greenish-white, speckled shells. In less than a month the young birds were able to fly quite well and collect their food.

Scarlet Tanager

(Piranga erythromelas) Tanager family

Called also: BLACK-WINGED REDBIRD ; FIREBIRD ; CAN-
ADA TANAGER ; POCKET-BIRD

Length—7 to 7.5 inches. About one-fourth smaller than the robin.
Male—In spring plumage: Brilliant scarlet, with black wings and
  tail. Under wing coverts grayish white. *In autumn:* Similar to female.
Female—Olive-green above; wings and tail dark, lightly margined
  with olive. Underneath greenish yellow.
Range—North America to northern Canada boundaries, and southward in winter to South America.
Migrations—May. October. Summer resident.

The gorgeous coloring of the scarlet tanager has been its
snare and destruction. The densest evergreens could not altogether hide this blazing target for the sportsman's gun, too often
fired at the instigation of city milliners. "Fine feathers make
fine birds"—and cruel, silly women, the adage might be adapted
for latter-day use. This rarely beautiful tanager, thanks to them,
is now only an infrequent flash of beauty in our country roads.

Instinct leads it to be chary of its charms; and whereas it
used to be one of the commonest of bird neighbors, it is now shy
and solitary. An ideal resort for it is a grove of oak or swamp
maple near a stream or pond where it can bathe. Evergreen
trees, too, are favorites, possibly because the bird knows how
exquisitely its bright scarlet coat is set off by their dark background.

High in the tree-tops he perches, all unsuspected by the visitor passing through the woods below, until a burst of rich, sweet
melody directs the opera-glasses suddenly upward. There we
detect him carolling loud and cheerfully, like a robin. He is an
apparition of beauty—a veritable bird of paradise, as, indeed, he
is sometimes called. Because of their similar coloring, the tanager and cardinal are sometimes confounded, but an instant's
comparison of the two birds shows nothing in common except
red feathers, and even those of quite different shades. The inconspicuous olive-green and yellow of the female tanager's plumage
is another striking instance of Nature's unequal distribution of

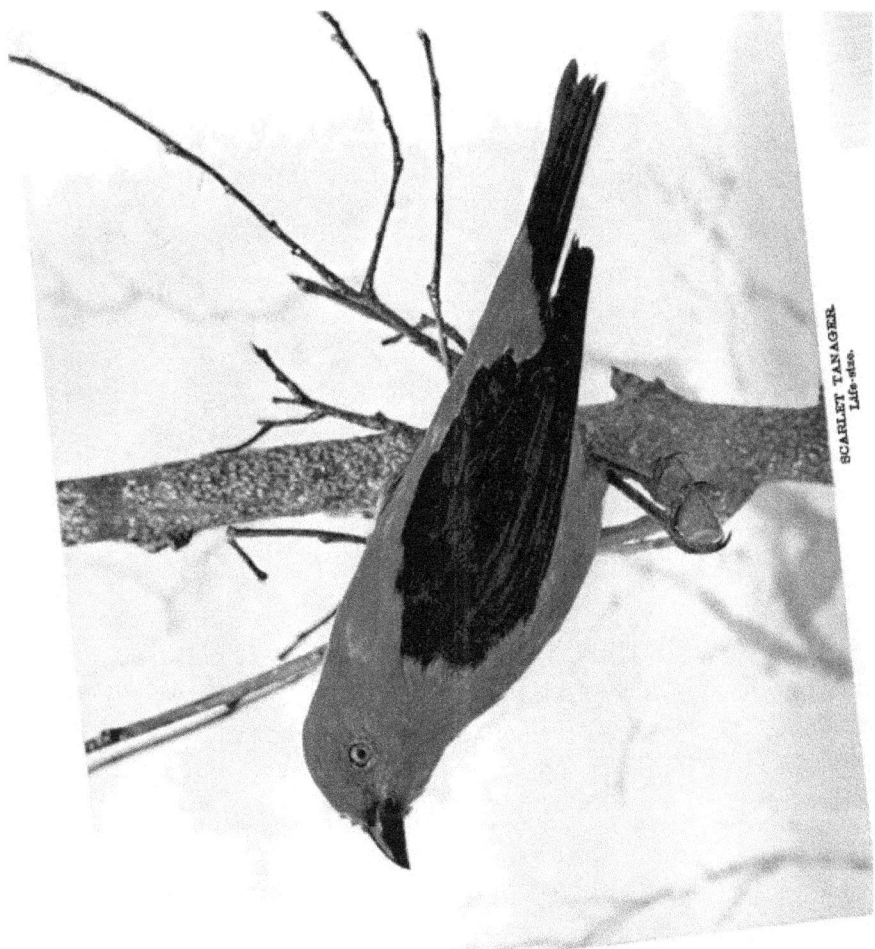

gifts; but if our bright-colored birds have become shockingly few under existing conditions, would any at all remain were the females prominent, like the males, as they brood upon the nest? Both tanagers construct a rather disorderly-looking nest of fibres and sticks, through which daylight can be seen where it rests securely upon the horizontal branch of some oak or pine tree; but as soon as three or four bluish-green eggs have been laid in the cradle, off goes the father, wearing his tell-tale coat, to a distant tree. There he sings his sweetest carol to the patient, brooding mate, returning to her side only long enough to feed her with the insects and berries that form their food.

Happily for the young birds' fate, they are clothed at first in motley, dull colors, with here and there only a bright touch of scarlet, yellow, and olive to prove their claim to the parent whose gorgeous plumage must be their admiration. But after the moulting season it would be a wise tanager that knew its own father. His scarlet feathers are now replaced by an autumn coat of olive and yellow not unlike his mate's.

Pine Grosbeak

(Pinicola enucleator) Finch family

Called also : PINE BULLFINCH

Length—Variously recorded from 6.5 to 11 inches. Specimen measured 8.5 inches. About one-fifth smaller than the robin.

Male—General color strawberry-red, with some slate-gray fleckings about head, under wings, and on legs. Tail brown; wings brown, marked with black and white and slate. A band-shaped series of markings between the shoulders. Underneath paler red, merging into grayish green. Heavy, conspicuous bill.

Female—Ash-brown. Head and hind neck yellowish brown, each feather having central dusky streak. Cheeks and throat yellowish. Beneath ash-gray, tinged with brownish yellow under tail.

Range—British American provinces and northern United States.

Migrations—Irregular winter visitors; length of visits as uncertain as their coming.

As inseparable as bees from flowers, so are these beautiful winter visitors from the evergreen woods, where their red feathers, shining against the dark-green background of the trees, give them charming prominence; but they also feed freely upon the buds of various deciduous trees.

South of Canada we may not look for them except in the severest winter weather. Even then their coming is not to be positively depended upon; but when their caprice—or was it an unusually fierce northern blast?—sends them over the Canada border, it is a simple matter to identify them when such brilliant birds are rare. The brownish-yellow and grayish females and young males, however, always seem to be in the majority with us, though our Canadian friends assure us of the irreproachable morals of this gay bird.

Wherever there are clusters of pine or cedar trees, when there is a flock of pine grosbeaks in the neighborhood, you may expect to find a pair of birds diligently feeding upon the seeds and berries. No cheerful note escapes them as they persistently gormandize, and, if the truth must be confessed, they appear to be rather stupid and uninteresting, albeit they visit us at a time when we are most inclined to rapture over our bird visitors. They are said to have a deliciously sweet song in the nesting season, when, however, few except the Canadian *voyageurs* hear it.

American Crossbill

(Loxia curvirostra minor) Finch family

Called also : RED CROSSBILL

Length—6 to 7 inches. About the size of the English sparrow.

Male—General color Indian red, passing into brownish gray, with red tinge beneath. Wings (without bands), also tail, brown. Beak crossed at the tip.

Female—General color greenish yellow, with brownish tints. Dull-yellowish tints on head, throat, breast, and underneath. Wings and tail pale brown. Beak crossed at tip.

Range—Pennsylvania to northern British America. West of Mississippi, range more southerly.

Migrations—Irregular winter visitor. November. Sometimes resident until April.

It is a rash statement to say that a bird is rare simply because you have never seen it in your neighborhood, for while you are

RED CROSSBILLS.
⅔ Life-size.

going out of the front door your *rara avis* may be eating the crumbs about your kitchen. Even with our eyes and ears constantly alert for some fresh bird excitement, our phlegmatic neighbor over the way may be enjoying a visit from a whole flock of the very bird we have been looking and listening for in vain all the year. The red crossbills are capricious little visitors, it is true, but by no means uncommon.

About the size of an English sparrow, of a brick or Indian red color, for the most part, the peculiarity of its parrot-like beak is its certain mark of identification.

Longfellow has rendered into verse the German legend of the crossbill, which tells that as the Saviour hung upon the cross, a little bird tried to pull out the nails that pierced His hands and feet, thus twisting its beak and staining its feathers with the blood.

At first glance the birds would seem to be hampered by their crossed beaks in getting at the seeds in the pine cones—a superficial criticism when the thoroughness and admirable dexterity of their work are better understood.

Various seeds of fruits, berries, and the buds of trees enlarge their bill of fare. They are said to be inordinately fond of salt. Mr. Romeyn B. Hough tells of a certain old ice-cream freezer that attracted flocks of crossbills one winter, as a salt-lick attracts deer. Whether the traditional salt that may have stuck to the bird's tail is responsible for its tameness is not related, but it is certain the crossbills, like most bird visitors from the far north, are remarkably gentle, friendly little birds. As they swing about the pine trees, parrot-fashion, with the help of their bill, calling out *kimp, kimp,* that sounds like the snapping of the pine cones on a sunny day, it often seems easily possible to catch them with the hand.

There is another species of crossbill, called the White-winged *(Loxia leucoptera),* that differs from the preceding chiefly in having two white bands across its wings and in being more rare.

The Redpoll

(Acanthis linaria) Finch family

Called also : REDPOLL LINNET; LITTLE SNOWBIRD; LESSER
REDPOLL

Length—5.25 to 5.5 inches.   About an inch shorter than the English sparrow.

Male—A rich crimson wash on head, neck, breast, and lower
back, that is sometimes only a pink when we see the bird in
midwinter.   Grayish-brown, sparrowy feathers show underneath the red wash.   Dusky wings and tail, the feathers
more or less edged with whitish.   Soiled white underneath;
the sides with dusky streaks.   Bill sharply pointed.

Female—More dingy than male, sides more heavily streaked, and
having crimson only on the crown.

Range—An arctic bird that descends irregularly into the northern
United States.

Migrations—An irregular winter visitor.

"Ere long, amid the cold and powdery snow, as it were a
fruit of the season, will come twittering a flock of delicate crim-
son-tinged birds, lesser redpolls, to sport and feed on the buds
just ripe for them on the sunny side of a wood, shaking down
the powdery snow there in their cheerful feeding, as if it were
high midsummer to them."   Thoreau's beautiful description of
these tiny winter visitors, which should be read entire, shows
the man in one of his most sympathetic, exalted moods, and it is
the best brief characterization of the redpoll that we have.

When the arctic cold becomes too cruel for even the snow-
birds and crossbills to withstand, flocks of the sociable little redpolls flying southward are the merest specks in the sullen, gray
sky, when they can be seen at all.   So high do they keep that
often they must pass above our heads without our knowing it.
First we see a quantity of tiny dots, like a shake of pepper, in the
cloud above, then the specks grow larger and larger, and finally
the birds seem to drop from the sky upon some tall tree that they
completely cover—a veritable cloudburst of birds.   Without
pausing to rest after the long journey, down they flutter into the
weedy pastures with much cheerful twittering, to feed upon
whatever seeds may be protruding through the snow.   Every

action of a flock seems to be concerted, as if some rigid disciplinarian had drilled them, and yet no leader can be distinguished in the merry company. When one flies, all fly; where one feeds, all feed, and by some subtle telepathy all rise at the identical instant from their feeding ground and cheerfully twitter in concert where they all alight at once. They are more easily disturbed than the goldfinches, that are often seen feeding with them in the lowlands; nevertheless, they quite often venture into our gardens and orchards, even in suburbs penetrated by the trolley-car.

Usually in winter we hear only their lisping call-note; but if the birds linger late enough in the spring, when their "fancy lightly turns to thoughts of love," a gleeful, canary-like song comes from the naked branches, and we may know by it that the flock will soon disappear for their nesting grounds in the northern forests.

The Greater Redpoll (*Acanthis linaria rostrata*) may be distinguished from the foregoing species by its slightly larger size, darker upper parts, and shorter, stouter bill. But the notes, habits, and general appearance of both redpolls are so nearly identical that the birds are usually mistaken for each other.

Purple Finch

(*Carpodacus purpureus*) Finch family

Called also : PURPLE LINNET

Length—6 to 6.25 inches. About the same size as the English sparrow.

Male—*Until two years old*, sparrow-like in appearance like the female, but with olive-yellow on chin and lower back. *Afterwards* entire body suffused with a bright raspberry-red, deepest on head, lower back, and breast, and other parts only faintly washed with this color. More brown on back; and wings and tail, which are dusky, have some reddish-brown feathers. Underneath grayish white. Bill heavy. Tail forked.

Female—Grayish olive-brown above; whitish below; finely streaked everywhere with very dark brown, like a sparrow. Sides of breast have arrow-shaped marks. Wings and tail darkest.

Range—North America, from Columbia River eastward to Atlantic, and from Mexico northward to Manitoba. Most common in Middle States and New England. Winters south of Pennsylvania.

Migrations—March. November. Common summer resident. Rarely individuals winter at the north.

In this "much be-sparrowed country" of ours familiarity is apt to breed contempt for any bird that looks sparrowy, in which case one of the most delicious songsters we have might easily be overlooked. It is not until the purple finch reaches maturity in his second year that his plumage takes on the raspberry-red tints that some ornithologists named purple. Oriental purple is our magenta, it is true, but not a raspberry shade. Before maturity, but for the yellow on his lower back and throat, he and his mate alike suggest a song-sparrow; and it is important to note their particularly heavy, rounded bills, with the tufts of feathers at the base, and their forked tails, to name them correctly. But the identification of the purple finch, after all, depends quite as much upon his song as his color. In March, when flocks of these birds come north, he has begun to sing a little ; by the beginning of May he is desperately in love, and sudden, joyous peals of music from the elm or evergreen trees on the lawn enliven the garden. How could his little brown lady-love fail to be impressed with a suitor so gayly dressed, so tender and solicitous, so deliciously sweet-voiced ? With fuller, richer song than the warbling vireo's, which Nuttall has said it resembles, a perfect ecstasy of love pours incessantly from his throat during the early summer days. There is a suggestion of the robin's love-song in his, but its copiousness, variety, and rapidity give it a character all its own.

In some old, neglected hedge or low tree about the country-place a flat, grassy nest, lined with horsehair, contains four or five green eggs in June, and the old birds are devotion itself to each other, and soon to their young, sparrowy brood.

But when parental duties are over, the finches leave our lawns and gardens to join flocks of their own kind in more remote orchards or woods, their favorite haunts. Their subdued warble may be heard during October and later, as if the birds were humming to themselves.

Much is said of their fondness for fruit blossoms and tree buds, but the truth is that noxious insects and seeds of grain

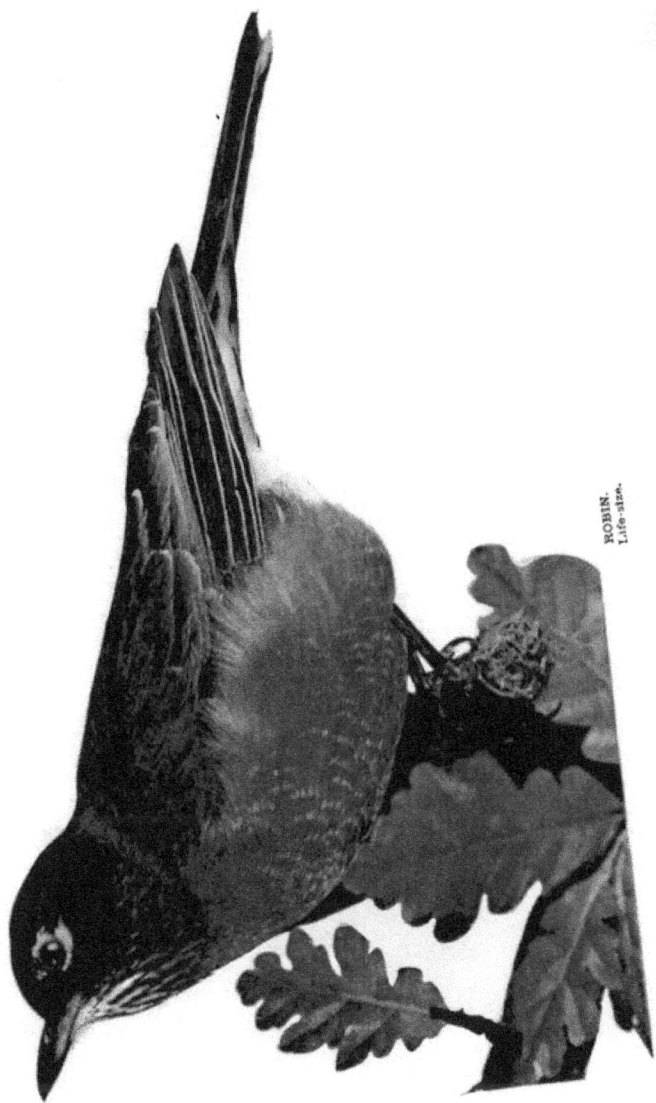

constitute their food in summer, the berries of evergreens in winter. To a bird so gay of color, charming of voice, social, and trustful of disposition, surely a few blossoms might be spared without grudging.

The American Robin

(Merula migratoria) Thrush family

Called also: RED-BREASTED OR MIGRATORY THRUSH ; ROBIN-REDBREAST

Length—10 inches.

Male—Dull brownish olive-gray above. Head black; tail brownish black, with exterior feathers white at inner tip. Wings dark brownish. Throat streaked with black and white. White eyelids. Entire breast bright rusty red; whitish below the tail.

Female—Duller and with paler breast, resembling the male in autumn.

Range—North America, from Mexico to arctic regions.

Migrations—March. October or November. Often resident throughout the year.

It seems almost superfluous to write a line of description about a bird that is as familiar as a chicken; yet how can this nearest of our bird neighbors be passed without a reference? Probably he was the very first bird we learned to call by name.

The early English colonists, who had doubtless been brought up, like the rest of us, on "The Babes in the Wood," named the bird after the only heroes in that melancholy tale; but in reality the American robin is a much larger bird than the English robin-redbreast and less brilliantly colored. John Burroughs calls him, of all our birds, "the most native and democratic."

How the robin dominates birddom with his strong, aggressive personality! His voice rings out strong and clear in the early morning chorus, and, more tenderly subdued at twilight, it still rises above all the sleepy notes about him. Whether lightly tripping over the lawn after the "early worm," or rising with his sharp, quick cry of alarm, when startled, to his nest near by, every motion is decided, alert, and free. No pensive hermit of the woods, like his cousins, the thrushes, is this joyous, vigorous "bird of the morning." Such a presence is inspiriting.

225

Does any bird excel the robin in the great variety of his vocal expressions ? Mr. Parkhurst, in his charming "Birds' Calendar," says he knows of "no other bird that is able to give so many shades of meaning to a single note, running through the entire gamut of its possible feelings. From the soft and mellow quality, almost as coaxing as a dove's note, with which it encourages its young when just out of the nest, the tone, with minute gradations, becomes more vehement, and then harsh and with quickened reiteration, until it expresses the greatest intensity of a bird's emotions. Love, contentment, anxiety, exultation, rage—what other bird can throw such multifarious meaning into its tone ? And herein the robin seems more nearly human than any of its kind."

There is no one thing that attracts more birds about the house that a drinking-dish—large enough for a bathtub as well; and certainly no bird delights in sprinkling the water over his back more than a robin, often aided in his ablutions by the spattering of the sparrows. But see to it that this drinking-dish is well raised above the reach of lurking cats.

While the robin is a famous splasher, his neatness stops there. A robin's nest is notoriously dirty within, and so carelessly constructed of weed-stalks, grass, and mud, that a heavy summer shower brings more robins' nests to the ground than we like to contemplate. The color of the eggs, as every one knows, has given their name to the tint. Four is the number of eggs laid, and two broods are often reared in the same nest.

Too much stress is laid on the mischief done by the robins in the cherry trees and strawberry patches, and too little upon the quantity of worms and insects they devour. Professor Treadwell, who experimented upon some young robins kept in captivity, learned that they ate sixty-eight earthworms daily—"that is, each bird ate forty-one per cent. more than its own weight in twelve hours! The length of these worms, if laid end to end, would be about fourteen feet. Man, at this rate, would eat about seventy pounds of flesh a day, and drink five or six gallons of water."

ORCHARD ORIOLE.
⅓ Life-size.

Orchard Oriole

(Icterus spurius) Blackbird and Oriole family

Called also : ORCHARD STARLING ; ORCHARD HANG-NEST

Length—7 to 7.3 inches. About one-fourth smaller than the robin.

Male—Head, throat, upper back, tail, and part of wings black. Breast, rump, shoulders, under wing and tail coverts, and under parts bright reddish brown. Whitish-yellow markings on a few tail and wing feathers.

Female—Head and upper parts olive, shading into brown; brighter on head and near tail. Back and wings dusky brown, with pale-buff shoulder-bars and edges of coverts. Throat black. Under parts olive, shading into yellow.

Range—Canada to Central America. Common in temperate latitudes of the United States.

Migrations—Early May. Middle of September. Common summer resident.

With a more southerly range than the Baltimore oriole and less conspicuous coloring, the orchard oriole is not so familiar a bird in many Northern States, where, nevertheless, it is quite common enough to be classed among our would-be intimates. The orchard is not always as close to the house as this bird cares to venture; he will pursue an insect even to the piazza vines.

His song, says John Burroughs, is like scarlet, "strong, intense, emphatic," but it is sweet and is more rapidly uttered than that of others of the family. It is ended for the season early in July.

This oriole, too, builds a beautiful nest, not often pendent like the Baltimore's, but securely placed in the fork of a sturdy fruit tree, at a moderate height, and woven with skill and precision, like a basket. When the dried grasses from one of these nests were stretched and measured, all were found to be very nearly the same length, showing to what pains the little weaver had gone to make the nest neat and pliable, yet strong. Four cloudy-white eggs with dark-brown spots are usually found in the nest in June.

INDEX

The figures in black-faced type indicate the page upon which the biography of the bird is given.

Index

Index

232

Index